HARDPRESS.NET
HOME OF HARD-TO-FIND BOOKS

Arctic Searching Expedition:
by Sir John Richardson

Arctic Searching Expedition:

John Richardson, John Franklin

93

PLATE VIII

ARC...

...CHEN...

...L OF...

SIR JOHN FRANKLIN

...JOHN RICHARDS...

IN...
VOL. I.

PUBLISH... BY AUTHORITY.

...
...OWN, GRE... AND LONGMANS.
...

ARCTIC
SEARCHING EXPEDITION:

A

JOURNAL OF A BOAT-VOYAGE

THROUGH RUPERT'S LAND AND THE ARCTIC SEA,

IN SEARCH OF

THE DISCOVERY SHIPS UNDER COMMAND OF

SIR JOHN FRANKLIN.

WITH AN APPENDIX ON THE PHYSICAL GEOGRAPHY
OF NORTH AMERICA.

BY SIR JOHN RICHARDSON, C.B., F.R.S.

INSPECTOR OF NAVAL HOSPITALS AND FLEETS,
ETC. ETC. ETC.

IN TWO VOLUMES.

VOL. I.

PUBLISHED BY AUTHORITY.

LONDON:
LONGMAN, BROWN, GREEN, AND LONGMANS.
1851.

203. a. 187.

NOTICE.

In the Indian names which occur in the following narrative, *u* is to be sounded like *oo*, in "moon;" *yu* as "yule," or like "you;" and *i* as in "ravine."

LONDON:
SPOTTISWOODES and SHAW.
New-street-Square.

CONTENTS.

CHAPTER I.

CHAP. II.

CHAP. III.

CHAP. IV.

CHAP. V.

CHAP. VI.

CHAP. VII.

CHAP. VIII.

CHAP. IX.

CHAP. X.

LIST OF PLATES.

WOODCUTS AND DIAGRAMS.

ERRATUM.

Page 35. line 3. from bottom, for " 1849 " read " 1848."

ARCTIC

SEARCHING EXPEDITION.

CHAPTER I.

ROUTE ASSIGNED TO THE EXPEDITION UNDER COMMAND OF SIR JOHN FRANKLIN.——NAMES OF THE OFFICERS.——EREBUS AND TERROR.——DATE OF ITS SAILING.——LAST LETTERS.—— SIR JOHN FRANKLIN'S LAST OFFICIAL LETTER.——LAST SIGHT OF THE EXPEDITION.——SIR JOHN ROSS PROPOSES A SEARCH. ——DISCUSSION OF VARIOUS OPINIONS OFFERED RESPECTING THE FATE OF THE EXPEDITION.——PLANS OF SEARCH ADOPTED. ——MAIN OBJECTS OF THE OVERLAND SEARCHING EXPEDITION. ——INSTRUCTIONS FROM THE ADMIRALTY.

HER MAJESTY's government having deemed it expedient that a further attempt should be made for the accomplishment of a north-west passage by sea from the Atlantic to the Pacific, the " Erebus " and " Terror " were fitted out for that service, and placed under the command of Captain Sir John Franklin, K. C. H. He was directed by the Admiralty instructions, dated on the 5th of May, 1845, to proceed with all despatch to Lancaster

VOL. I.   B

Sound, and, passing through it, to push on to the westward, in the latitude of 74½°, without loss of time or *stopping to examine any openings to the northward*, until he reached the longitude of Cape Walker, which is situated in about 98° west. He was to use every effort to penetrate to the *southward* and *westward* of that point, and to pursue as direct a course for Beering's Straits as circumstances might permit. He was cautioned not to attempt to pass by the western extremity of Melville Island, until he had ascertained that a permanent barrier of ice or other obstacle closed the prescribed route. In the event of not being able to penetrate to the westward, he was to enter Wellington Sound in his second summer.

He was further directed to transmit accounts of his proceedings to the Admiralty, by means of the natives and the Hudson's Bay Company, should opportunities offer; and also, after passing the 65th meridian, to throw overboard daily a copper cylinder, containing a paper stating the ship's position. It was also understood that he would cause piles of stones or signal-posts to be erected on conspicuous headlands at convenient times, though the instructions do not contain a clause to that effect.*

* The instructions are published at length in a parliamentary Blue Book, and all known particulars respecting the expedition have been communicated from time to time to the public

The following officers joined the expedition :—

EREBUS.

Captain, Sir John Franklin, Kt. K. C. H.
Commander, James Fitzjames.
Lieutenant, Graham Gore.
Lieut., H. P. D. Le Vesconte.
Lieut., James W. Fairholme.
Ice-Master, James Read.
Surgeon, Stephen S. Stanley.
Paymaster, C. H. Osmer.
Assist.-Surg., H. D. S. Goodsir.
Sec. Master, Henry F. Collins.

TERROR.

Capt., Francis R. M. Crozier.
Lieutenant, Edward Little.
Lieut., George H. Hodgson.
Lieutenant, John Irving.
Ice-Master, Thomas Blanky.
Surgeon, John S. Peddie.
Assist.-Surgeon, A. M'Donald.
Sec. Mr., Gillies A. Maclean.
Clerk-in-Charge, Edward J. H. Helpman.

And the conjoined crews of the two ships amounted to 130 souls.

The " Erebus," originally built for a bomb-vessel, and therefore strongly framed, was of 370 tons measurement, and had been fortified, in 1839, after the most approved plan, by an extra or double exterior planking and diagonal bracing within, for Sir James C. Ross's antarctic voyage, from which she returned in 1843. Having been carefully examined and refitted for Sir John Franklin, she was considered to be as strongly prepared to resist the pressure of the ice as the resources of science, and the utmost care of Mr. Rice, the skilful master-shipwright who superintended the preparations, could ensure. The " Terror," of 340 tons,

by the same channel. The above abstract mentions the leading points which would direct the course of the expedition.

was also constructed for a bomb-vessel, and had the bluff form, capacious hold, and strong framework of that class of war vessels. When commanded by Captain Sir George Back, on his voyage to Repulse Bay in 1836–7, she had been beset for more than eleven months in drifting floes of ice, and exposed to every variety of assault and pressure to which a vessel was liable in such a dangerous position. In this severe and lengthened trial, the "Terror" had been often pressed more or less out of the water, or thrown over on one side, and had, in consequence thereof, sustained some damage, particularly in the stern post. All defects, however, were made good in 1839, when she sailed for the Antarctic Seas, under the command of Captain Crozier, the second officer of Sir James C. Ross's expedition. She was again examined, and made as strong as ever, before Captain Crozier took the command of her a second time in 1845.

The best plans that former experience could suggest for ventilating and warming the ships in the winter were adopted, and full supplies of every requisite for arctic navigation were provided, including an ample stock of warm bedding, clothing, and provisions, with a proportion of preserved meats and pemican.

The expedition sailed from England on the 19th of May, 1845, and, early in July, had reached Whalefish Islands, near Disco, on the Greenland coast of Davis's Straits, where, having found a convenient port, the transport which accompanied it was cleared and sent home to England, bringing the last letters that have been received from the officers or crew. The following extract of a letter, from Lieutenant Fairholme, of the " Erebus," will serve to show the cheerful anticipation of success which prevailed throughout the party, and the happy terms on which they were with each other : —

" We have anchored in a narrow channel between two of the islands, protected on all sides by land, and in as convenient a place for our purpose as could possibly be found. Here we are with the transport lashed alongside, transferring most actively all her stores to the two ships. I hope that this operation will be completed by to-morrow night, in which case Wednesday will be devoted to swinging the ships for local attraction, and I suppose Thursday will see us under way with our heads to the northward. We have had the observatory up here, on a small rock on which Parry formerly observed, and have got a very satisfactory set of magnetic and other observations. Of our prospects we know little more than when we left England, but look forward with anxiety to our reaching 72°, where it seems we are likely to meet the first obstruction, if any exists. On board we are as comfortable as it is possible to be. I need hardly tell you how much we are all delighted with our Captain. He has, I am sure, won not only the respect but the love of every person on board by his amiable manner and kindness to all; and his influence is always employed for some good purpose both among the officers and men. He has been most successful in his selection of officers, and a more agreeable set could hardly be found. Sir John is in much better health than when we left England, and really looks ten years younger. He takes an active part in every thing that goes on, and his long experience in such services as this makes him a most valuable adviser. *July* 10*th*.——The transport is just reported clear, so I hope that we may be able to swing the ships to-morrow and get away on Saturday. We are very much crowded; in fact, not an inch of stowage has been lost, and the decks are still covered with casks, &c. Our supply of coals has encroached seriously on the ship's stowage; but as we consume both this and provisions as we go, the evil will be continually lessening."

Letters from most of the other officers, written in a similarly buoyant and hopeful spirit, were received in England at the same time with the above. An extract of a letter from Sir John

Franklin himself to Lieutenant Colonel Sabine deserves to be quoted, as expressing his own opinion of his resources, and also his intention of remaining out should he fail after a second winter in finding an outlet to the south-westward from Barrow's Strait. The letter is dated from Whalefish Islands, on the 9th of July, 1845, and, after noticing that the " Erebus " and " Terror " had on board provisions, fuel, clothing, and stores for three years complete, from that date, adds, " I hope my dear wife and daughter will not be over anxious if we should not return by the time they have fixed upon ; and I must beg of you to give them the benefit of your advice and experience when that time arrives, for you know well that, without success in our object, even after the *second winter*, we should wish to try some other channel if the state of our provisions and the health of the crews justify it."

The following is the last official letter written by Sir John Franklin to the Admiralty.

> " Her Majesty's Ship ' Erebus,'
> Whalefish Islands, July 12. 1845.

" Sir,
" I have the honour to acquaint you, for the information of the Lords Commissioners of the Admiralty, that Her Majesty's ships ' Erebus ' and ' Terror,' with the transport, arrived at this anchorage on the 4th instant, having had a passage of one month from Stromness. The transport was immediately taken alongside this ship, that

she might be more readily cleared; and we have been constantly employed at that operation till last evening, the delay having been caused not so much in getting the stores transferred to either of the ships, as in making the best stowage of them below, as well as on the upper deck. The ships are now complete with supplies of every kind for three years: they are, therefore, very deep; but happily we have no reason to expect much sea as we proceed further.

" The magnetic instruments were landed the same morning; so also were the other instruments requisite for ascertaining the position of the observatory; and it is satisfactory to find that the results of the observations for latitude and longitude accord very nearly with those assigned to the same place by Sir Edward Parry. Those for dip and variation are equally satisfactory, which were made by Captain Crozier with the instruments belonging to the ' Terror,' and by Commander James with those of the ' Erebus.'

" The ships are now being swung, for the purpose of ascertaining the dip and deviation of the needle on board, as was done at Greenhithe; which I trust will be completed this afternoon, and I hope to be able to sail in the night.

" The governor and principal persons are at this time absent from Disco; so that I have not been able to receive any communication from head quarters as to the state of the ice to the north. I have, however, learned from a Danish carpenter in charge of the Esquimaux at these islands, that, though the winter was severe, the spring was not later than usual, nor was the ice later in breaking away hereabout. He supposes, also, that it is now loose as far as 74°, and as far as Lancaster Sound, without much obstruction.

" The transport will sail for England this day.   I shall
instruct the agent, Lieutenant Griffiths, to proceed to
Deptford, and report his arrival to the Secretary of the
Admiralty.   I have much satisfaction in bearing my testimony to the careful and zealous manner in which Lieutenant Griffiths has performed the service entrusted to
him, and would beg to recommend him, as an officer who
appears to have seen much service, to the favourable consideration of their Lordships.

" It is unnecessary to assure their Lordships of the
energy and zeal of Captain Crozier, Commander Fitzjames, and of the officers and men with whom I have the
happiness of being employed on this service.

" I have, &c.

" JOHN FRANKLIN,

" Captain.

" The Right Hon. H. L. Cary, M. P.
    &c.   &c.   &c."

The two ships were seen on the 26th of the same
month (July) in latitude 74° 48' N., longitude 66°
13' W., moored to an iceberg, waiting for a favourable opportunity of entering or rounding the
" middle ice " and crossing to Lancaster Sound,
distant in a direct westerly line from their position
about 220 geographical miles.   On that day a boat
from the discovery ships, manned by seven officers,
one of whom was Commander Fitzjames, boarded
the " Prince of Wales," whaler, Captain Dannett.
They were all in high spirits, and invited Captain
Dannett to dine with Sir John Franklin on the

following day, which had he done, he would doubt-less have been the bearer of letters for England, but a favourable breeze springing up he separated from them. The ice was then heavy but loose, and the officers expressed good hopes of soon ac-complishing the enterprise. Captain Dannett was favoured with very fine weather during the three following weeks, and thought that the expedition must have made good progress. This was the last sight that was obtained of Franklin's ships.

In January 1847, a year and a half after the above date, Captain Sir John Ross addressed a letter to the Admiralty, wherein he stated his conviction that the discovery ships were frozen up at the *western end of Melville Island*, from whence their return would be for ever prevented by the accu-mulation of ice behind them, and volunteered his services to carry relief to the crews. Sir John also laid statements of his apprehensions before the Royal and Geographical Societies, and, the public attention being thereby roused, several writers in the newspapers and other periodicals published their sentiments on the subject, a variety of plans of relief were suggested, and many volunteers came forward to execute them.

The Lords Commissioners of the Admiralty, though judging that the second winter was too early a period of Sir John Franklin's absence to give rise to well founded apprehensions for his safety, lost no time in calling for the opinions of several naval officers who were well acquainted with arctic navigation, and in concerting plans of relief, to be carried out when the proper time should arrive.

A brief review of the replies most worthy of notice may help the reader to form a judgment of the plans that were eventually adopted by the Admiralty for the discovery and relief of the absent voyagers. It is convenient to consider first the notions of those who believe that Sir John Franklin never entered Lancaster Sound, either because the ships met with some fatal disaster in Baffin's Bay, and went down with the entire loss of both crews, or that Sir John endeavoured to fulfil the purposes of the expedition by taking some other route than the one exclusively marked out for him by his instructions. That the ships were not suddenly wrecked by a storm, or overwhelmed by the pressure of the ice, may be concluded from facts gathered from the records of the Davis's Straits whale-fishery, by which we learn that of the many vessels which have been crushed in the ice, in the course of several centuries, the whole or greater part of the crews have almost always escaped with their boats. It is, therefore, scarcely possible to believe that two vessels so strongly fortified as the "Erebus" and "Terror," and found by previous trials to be capable of sustaining so enormous a pressure, should both of them have been so suddenly crushed as to allow no time for active officers and men, disciplined and prepared for emergencies of the kind, to get out their boats. And having done so, they would have had little difficulty in reaching one of the many whalers, that were occupied in the pursuit of fish in those seas for six weeks after the discovery ships were last seen. Moreover, had the ships been wrecked, some fragments of their spars or hulls would have been found floating by the whalers, or being cast on the eastern or western shores of the bay, would have been reported by the Greenlanders or Eskimos. Neither are any severe storms recorded as having occurred then or there, nor did any unusual calamity befall the fishing vessels that year.

With respect to Sir John Franklin having chosen to enter Jones's or Smith's Sounds in preference to Lancaster Sound, his known habit of strict adherence to his instructions is a sufficient answer, and the extract quoted above from his letter to Lieutenant Colonel Sabine, which gives his latest thoughts on the subject, plainly says that such a course would not be pursued until a *second winter* had proved the impracticability of the route laid down for him. This point is mooted, because Mr. Hamilton, surgeon in Orkney, states that Sir John, when dining with him on the last day that he passed in Great Britain, mentioned his determination of trying Jones's Sound. But Sir John's communication to Colonel Sabine shows that this could be meant to refer only to the contingency of a full trial by Lancaster Sound proving fruitless. Supposing that, contrary to all former experience, he had found the mouth of Lancaster Sound so barred by ice as to preclude his entrance, then, after waiting till he had become convinced that it would remain closed for the season, he might have tried to find a way, by Jones's Sound, into Wellington Sound ; but in such a case, we may hold it as certain that he would have erected conspicuous cairns, and deposited memoranda of his past proceedings and future intentions, at the entrance of Lancaster Sound.

Taking it, then, for granted that the expedition entered Lancaster Sound, the most probable conjecture respecting the direction in which it advanced is that Sir John, literally following his instructions, did not stop to examine any openings either to the northward or southward of Barrow's Strait, but continued to push on to the westward until he reached Cape Walker in longitude 98°, when he inclined to the south-west, and steered as directly as he could for Beering's Straits. But even supposing that the state of the ice permitted him to take the desired route, and to turn to the south-westward by the first opening beyond the 98th meridian, we are ignorant of the exact position of that opening, the tract between Cape Walker and Banks's Land being totally unknown. That a passage to the southward does exist in that space, and terminates between Victoria and Wollaston Lands in Coronation Gulf, is inferred from the observed setting of the flood tide. There is, it is true, an uncertainty in our endeavours to determine the directions of the tides in these narrow seas, where the currents are influenced by prevailing winds; but Mr. Thomas Simpson, who was an acute observer, remarked that the flood tide brought much ice into Coronation Gulf round the west end of Victoria Land, and facts collected on three visits which I have made to that gulf lead me to concur with him. Entirely in accordance with this opinion is the fact noted by Sir Edward Parry, that the flood tide came from the north between Cornwallis and the neighbouring islands, and that the ice was continually setting round the west end of Melville Island and passing onwards to the south-east.

These observations, while they point to an opening to the eastward of Banks's Land, may be adduced as an argument against the existence of a passage directly to the westward between it and Melville Island; and, though they are not conclusive, they are supported by another remark of Sir Edward Parry's, that he thought there was some peculiar obstruction immediately to the west of that island, which produced a permanent barrier of ice.

But wherever the opening which we presume to exist may be situated, the channels among the islands are probably not direct, and may be intricate. Vessels, therefore, having pushed into one of them would be exposed to the ice closing in behind and barring all regress. Sir John Ross, whose opinions are first recorded in the parliamentary Blue Book, believes that "Sir John Franklin put his ships into the drift ice at the western end of Melville Island," and that, "if not totally lost, they must have been carried by the ice, which is known to drift to the southward, on land (Banks's Land) seen at a great distance in that direction, and from which the accumulation of ice behind them will," says he, "as in my own case, for ever prevent the return of the ships."

Sir W. Edward Parry is of opinion that Sir John Franklin would endeavour "to get to the southward and westward before he approached the south-western extremity of Melville Island, that is, between the 100th and 110th degree of longitude:"

"how far they may have penetrated to the southward between those meridians, must be a matter of speculation, depending on the state of the ice and the existence of land in a space hitherto blank in our maps." "Be this as it may, I (Sir W. E. Parry) consider it not improbable, as suggested by Dr. King, that an attempt will be made by them to fall back on the western coast of North Somerset, wherever that may be found, as being the nearest point affording a hope of communication, either with whalers or with ships sent expressly in search of the expedition."

Sir James C. Ross says: "It is far more probable, however, that Sir John Franklin, in obedience to his instructions, would endeavour to push the ships to the south and west as soon as they passed Cape Walker; and the consequence of such a measure, owing to the known prevalence of westerly winds, and the drift of the main body of the ice, would be, their inevitable embarrassment; and if he persevered in that direction, which he probably would do, I have no hesitation in stating my conviction, that he would never be able to extricate his ships, and would ultimately be obliged to abandon them. It is, therefore, in latitude 73° N. and longitude 135° W. that we may expect to find them involved in the ice, or shut up in some harbour."

The opinions here quoted are contingent on the supposition, that Sir John Franklin found the state of the ice to be such that he could take the routes in question ; but the several officers quoted admit that, in the event of no opening through the ice in a westerly or south-westerly direction being found, Sir John would attempt Wellington Sound, or any other northern opening that was accessible. Commander Fitzjames, in a letter dated January, 1845, says : " The north-west passage is certainly to be gone through by Barrow's Straits, but whether south or north of Parry's Group remains to be proved. I am for going far north, edging north-west till in longitude 140° W., if possible." Mr. John Barrow, to whom this letter was addressed, appends to it the following memorandum : " Captain Fitzjames was much inclined to try the passage to the northward of Parry's Islands, and he would no doubt endeavour to persuade Sir John Franklin to pursue that course, if they failed to get to the southward."

My own opinion, submitted to the Admiralty in compliance with their commands, was substantially the same with that of Sir James Clark Ross, though formed independently ; and I further suggested that, in the event of accident to the ships, or their abandonment in the ice, the members of the ex-

C

pedition would make either for Lancaster Sound to meet the whalers, or Mackenzie River to seek relief at the Hudson's Bay posts, as they judged either of these places most easy of attainment.

After deliberately weighing these and other suggestions, and fully considering the numerous plans submitted to them, the Admiralty determined that, if no intelligence of the missing ships arrived by the close of autumn, 1847, they would send out three several searching expeditions — one to Lancaster Sound, another down the Mackenzie River, and the third to Beering's Straits.

The object of the first, and the most important of the three, was to follow up the route supposed to have been pursued by Sir John Franklin; and, by searching diligently for any signal-posts he might have erected, to trace him out, and carry the required relief to his exhausted crews. Sir James Clark Ross was appointed to the command of this expedition, consisting of the " Enterprise " and " Investigator;" and, as his plan of proceeding bears upon my own instructions, I give it at length : —

" As vessels destined to follow the track of the expedition must necessarily encounter the same difficulties, and be liable to the same severe pressure from the great body of the ice they must pass through in their way to Lancaster Sound, it is desirable that two ships, of not less than 500 tons, be purchased for this service, and fortified and equipped, in every respect as were the 'Erebus' and 'Terror,' for the Antarctic Seas.

"Each ship should, in addition, be supplied with a small vessel or launch of about 20 tons, which she could hoist in, to be fitted with a steam-engine and boiler of ten-horse power, for a purpose to be hereafter noticed.

"The ships should sail at the end of April next, and proceed to Lancaster Sound, with as little delay as possible, carefully searching both shores of that extensive inlet, and of Barrow's Strait, and then progress to the westward.

"Should the period at which they arrive in Barrow's Strait admit of it, Wellington Channel should next be examined, and the coast between Cape Clarence and Cape Walker explored, either in the ships or by boats, as may at the time appear most advisable. As this coast has been generally found encumbered with ice, it is not desirable that both ships should proceed so far along it as to hazard their getting beset there and shut up for the winter; but in the event of finding a convenient harbour near Garnier Bay or Cape Rennell, it would be a good position in which to secure one of the ships for the winter.

"From this position the coast line might be explored, as far as it extends to the westward, by detached parties early in the spring, as well as the western coast of Boothia, a considerable distance to the southward; and at a more advanced period of the season the whole distance to Cape Nicolai might be completed.

"A second party might be sent to the south-west as far as practicable, and a third to the north-west, or in any other direction deemed advisable at the time.

"As soon as the formation of water along the coast

between the land and main body of the ice admitted, the
small steam-launch should be despatched into Lancaster
Sound, to communicate with the whale ships at the usual
time of their arrival in those regions, by which means
information of the safety or return of Sir John Franklin
might be conveyed to the ships before their liberation
from their winter quarters, as well as any further instructions the Lords Commissioners might be pleased to
send for their future guidance.

" The easternmost vessel having been safely secured in
winter quarters, the other ship should proceed alone to
the westward, and endeavour to reach Winter Harbour in
Melville Island, or some convenient port in Banks's Land,
in which to pass the winter.

" From this point, also, parties should be despatched
early in spring, before the breaking up of the ice.   The
first should trace the western coast of Banks's Land, and,
proceeding to Cape Bathurst, or some other conspicuous
point of the continent, previously agreed on with Sir John
Richardson, reach the Hudson's Bay Company's settlement of Fort Good Hope on the Mackenzie, whence they
may travel southward by the usual route of the traders
to York Factory, and thence to England.

" The second party should explore the eastern shore of
Banks's Land, and, making for Cape Krusenstern, communicate with Sir John Richardson's party on its descending
the Coppermine River, and either assist him in completing
the examination of Wollaston and Victoria Land, or
return to England by any route he should direct.

" These two parties would pass over that space in
which most probably the ships have become involved (if
at all), and would, therefore, have the best chance of
communicating to Sir John Franklin information of the

measures that have been adopted for his relief, and of directing him to the best point to proceed, if he should consider it necessary to abandon his ships.

" Other parties may be despatched, as might appear desirable to the commander of the expedition, according to circumstances ; but the steam-launches should certainly be employed to keep up the communication between the ships, to transmit such information for the guidance of each other as might be necessary for the safety and success of the undertaking.

(Signed)   " JAMES C. ROSS,

" Captain, R. N.

" Athenæum, 2 December, 1847."

By a subsequent arrangement between Sir James Ross and myself, under the sanction of the Admiralty, I undertook to deposit pemican at Fort Good Hope and Point Separation on the Mackenzie, and Capes Bathurst, Parry, Krusenstern, and Hearne, on the sea-coast, for the use of Sir James Ross's detached parties.

The Beering's Straits expedition was composed of the " Herald," Captain Kellet, then employed in surveying the Pacific coasts of America, and the " Plover," Commander Moore. The vessels were expected to arrive in Beering's Straits about the 1st of July, 1848, and were directed to " proceed along the American coast as far as possible, consistent with the certainty of preventing the ships being beset by the ice." A harbour was to be

c 3

sought for the "Plover" within the Straits, to which that vessel was to be conducted; and two whale-boats were to go on to the eastward in search of the missing voyagers, and to communicate, if possible, with the Mackenzie River party. The "Plover" was fitted out in the Thames in December, 1847; but having been found to leak when she went to sea, was compelled to put into Plymouth for repair, and did not finally leave England until February, 1848. This tardy departure, conjoined with her dull sailing, prevented her from passing Beering's Straits at all in 1848; but she wintered near Cape Tschukotskoi, on the Asiatic coast, just outside of the Straits.

The "Herald" visited Kotzebue Sound, repassed the Straits before the arrival of the "Plover," and returned to winter in South America, with the intention of going northwards again next season.

The main object of the searching party entrusted to my charge was to trace the coast between the Mackenzie and Coppermine Rivers, and the shores of Victoria and Wollaston Lands lying opposite to Cape Krusenstern. In a preceding page I have adduced reasons for believing that there is a passage to the northwards between these lands; and if so, its position makes it the most direct route from the continent to the unknown tract interposed between Cape Walker and Banks's Land, into which Sir John Franklin was expressly ordered to carry his ships. Should he have done so, and his egress by the way he entered be barred by the ice closing in behind him as already suggested, there remained a probability that the annual progression of the ice southwards would eventually carry the ships into Coronation Gulf, or, if abandoned before that event, their crews were to be sought for on their way to the continent.

At the time when Sir John Franklin left England, two other openings from the north into the sea washing the continental shores were supposed to exist. The most westerly of these is between Boothia and Victoria Land, and it was part of Sir James Ross's plan to examine the whole western side of Boothia and North Somerset by one of his steam-barges.

The other supposed entrance was by Regent's Inlet. Dease and Simpson had left only a small space unsurveyed between that inlet and the sea, which was known to afford in good seasons a passage all the way to Beering's Straits; and this might have recommended the route by Regent's Inlet for trial. But, exclusive of its being absolutely prohibited by Sir John Franklin's instructions, Sir Edward Parry and Sir James Ross, on whose opinions Sir John placed deservedly the greatest reliance, were decidedly averse to his attempting a passage in that direction; and it was known that Sir John Franklin had resolved on trying all the other openings before he entered Regent's Inlet, which was to be his last resource. It fortunately happened before any of the searching expeditions were finally organised, that the non-existence of a passage through that inlet was fully ascertained.

Mr. John Rae, a Chief Trader in the service of the Hudson's Bay Company, left Fort Churchill in the beginning of the summer of 1846, with two boats, for the express purpose of completing the survey of Regent's Inlet. He arrived in Repulse Bay in the month of August of that year, and immediately crossed an isthmus, forty-three miles wide, to the inlet, taking one boat with him. Finding that the season was too far advanced for him to complete the survey that year, he determined, with a boldness and confidence in his own resources that has never been surpassed, to winter in Repulse Bay, and to finish his survey of Regent's Inlet on the ice next spring; so that he might be able to return to Churchill and York Factory by open water in the summer of 1847. He therefore recrossed the isthmus again with his boat, and set about collecting provisions and fuel for a ten months' winter. To one less experienced and hardy, the desolate shores of Repulse Bay would have forbidden such an attempt. They yielded neither drift-wood nor shrubby plants of any kind; but Mr. Rae employed part of his men to gather the withered stems of the *Andromeda tetragona*, a small herbaceous plant which grew in abundance on the rocks, and to pile it in cocks like hay: others he set to build a house of stone and earth, large enough to shelter his party, amounting in all to sixteen; whilst he himself and his Eskimo interpreter were occupied in killing deer for winter consumption. He succeeded in laying up a sufficient stock of venison, and kept his people in health and strength for next year's operations, though not in comfort, for the chimney was so badly constructed for ventilation, that when the fire was lighted it was necessary to open the door, and thus to reduce the temperature of the apartment, nearly to that of the external air. The fire was, therefore, used as seldom as possible, and only for cooking or melting snow to drink. In the spring he completed the survey of Prince Regent's Inlet on foot, thereby proving that no passage existed through it, and confirming the Eskimo report, first made to Sir Edward Parry and afterwards to Sir John Ross. A party of Eskimo, who resided near Mr. Rae in the winter, informed him, through his interpreter, that they had not seen Franklin's ships, thereby excluding the Gulf of Boothia from the list of places to be searched.

Having thus mentioned the opinions most worthy of note, respecting the quarters in which search was to be made, the plans of search adopted by the Admiralty after duly weighing a great variety of suggestions, and the extent of coast and parts of the Arctic Sea embraced in the three expeditions of the summer of 1848, I subjoin the instructions I received from the Admiralty.

Instructions to Sir John Richardson, M. D., 16th March 1848. By the Commissioners for executing the office of Lord High Admiral, &c.

" Whereas we think fit that you should be employed on an overland expedition in search of Her Majesty's ships ' Erebus ' and ' Terror,' under the command of Captain Sir John Franklin, which ships are engaged in a voyage of discovery in the Arctic seas, you are hereby required and directed to take under your orders Mr. Rae, who has been selected to accompany you, and to leave England on the 25th instant by the mail steamer for Halifax in Nova Scotia, and New York; and on your arrival at the latter place, you are to proceed immediately to Montreal, for the purpose of conferring with Sir George Simpson, Governor of the Hudson's Bay Company's Settlements, and making arrangements with him for your future supplies and communications.

" You should next travel to Penetanguishene, on Lake Huron, and from thence, by a steamer, which sails on the 1st and 15th of every month of open water, to Saut Ste. Marie, at the foot of Lake Superior, and there embark in a canoe, which, with its crew, will have been provided for you, by that time, by Sir George Simpson.

" Following the usual canoe route by Fort William, Rainy Lake, the Lake of the Woods, Lake Winipeg, and the Saskatchewan River, it is hoped that you will overtake the boats now under charge of Mr. Bell, in July 1848, somewhere near Isle à la Crosse, or perhaps the Methy Portage.

" You will then send the canoe with its crew back to Canada, and having stowed the four boats for their sea voyage, you will go on as rapidly as you can to the mouth of the Mackenzie; leaving Mr. Bell to follow with the heavier laden barge, to turn off at Great Bear Lake, and erect your winter residence at Fort Confidence, establish fisheries, and send out hunters.

" Making a moderate allowance for unavoidable detention by ice, thick fogs, and storms, the examination of the coast between the Mackenzie and the Coppermine Rivers will probably occupy 30 days; but you cannot calculate to be able to keep the sea later than the 15th of September, for, from the beginning of that month, the young ice covers the sea almost every night, and very greatly impedes the boats, until the day is well advanced.

" If you reach the sea in the first week of August, it is hoped you will be able to make the complete voyage to the Coppermine River, and also to coast a considerable part of the western and southern shores of Wollaston Land, and to ascend the Coppermine to some convenient point, where the boats can be left with the provisions

ready for the next year's voyage; and you will instruct
Mr. Bell to send two hunters to the banks of the river
to provide food for the party on the route to Fort Confidence, and thus spare you any further consumption of
the pemican reserved for the following summer.

" As it may happen, however, from your late arrival on
the coast, or subsequent unexpected detentions, that you
cannot with safety attempt to reach the Coppermine, you
have our full permission in such a case to return to Fort
Good Hope, on the Mackenzie, there to deposit two of
the boats, with all the sea stores, and to proceed with the
other two boats, and the whole of the crews, to winter
quarters on Great Bear Lake.

" And you have also our permission to deviate from the
line of route along the coast, should you receive accounts
from the Eskimos, which may appear credible, of the
crews of the ' Erebus ' and ' Terror,' or some part of
them, being in some other direction.

" For the purpose of more widely extending your
search, you are at liberty to leave Mr. Rae and a party of
volunteers to winter on the coast, if, by the establishment
of a sufficient fishery, or by killing a number of deer or
musk oxen, you may be able to lay up provisions enough
for them until you can rejoin them next summer.

" As you have been informed by Captain Sir James
Ross, of Her Majesty's ship ' Enterprize,' who is about
to be employed on a similar search in another direction,
of the probable directions in which the parties he will
send out towards the continent will travel, you are to
leave a deposit of pemican for their use at the following
points—namely, Point Separation, Cape Bathurst, Cape
Parry, and Cape Krusenstern; and as Sir James Ross is
desirous that some pemican should be stored at Fort Good

Hope, for the use of a party which he purposes sending thither in the spring of 1849, you are to make the necessary arrangements with Sir George Simpson for that purpose, as his directions to that effect must be sent early enough to meet the Company's brigade of Mackenzie River boats at Methy Portage, in July 1848.

" Should it appear necessary to continue the search a second summer (1849), and should the boats have been housed on the Coppermine, you are to descend that river on the breaking up of the ice in June 1849, and to examine the passages between Wollaston and Banks's and Victoria Lands, so as to cross the routes of some of Sir James C. Ross's detached parties, and to return to Great Bear Lake in September 1849, and withdraw the whole party from thence to winter on Great Slave Lake, which would be as far south as you will have a prospect of travelling before the close of the river navigation.

" Should you have found it necessary to return to the Mackenzie (September 1848), instead of pushing on to the Coppermine, the search in the summer of 1849 would, of course, have to be commenced from the former river again; but should circumstances render it practicable and desirable to send some of the party down the Coppermine with one or two boats, you are at liberty to do so.

" A passage for yourself and Mr. Rae will be provided in the 'America,' British and North American mail-steamer, which sails from Liverpool on the 25th of March, and you will receive a letter of credit on Her Majesty's Consul at New York for the amount of the expense of your journey from New York to Saut Ste. Marie, and the carriage of the instruments, &c.

" And in the event of intelligence of the ' Erebus ' and ' Terror ' reaching England after your departure, a communication will be made to the Hudson's Bay Company to ascertain the most expeditious route to forward your recal.

" We consider it scarcely necessary to furnish you with any instructions contingent on a successful search after the above-mentioned expedition, or any parties belonging to it. The circumstances of the case, and your own local knowledge and experience, will best point out the means to be adopted for the speedy transmission to this country of intelligence to the above effect, as well as of aiding and directing in the return of any such parties to England.

" We are only anxious that the search so laudably undertaken by you and your colleagues should not be unnecessarily or hazardously prolonged; and whilst we are confident that no pains or labour will be spared in the execution of this service, we fear lest the zeal and anxiety of the party so employed may carry them further than would be otherwise prudent.

" It is on this account you are to understand that your search is not to be prolonged after the winter of 1849, and which will be past on the Great Slave Lake; but that, at the earliest practicable moment after the breaking up of the weather in the spring of 1850, you will take such steps for the return of the party under your orders to England as circumstances may render expedient.

" It must be supposed that the instructions now afforded you can scarcely meet every contingency that may arise out of a service of the above description; but reposing, as we do, the utmost confidence in your discretion and judgment, you are not only at liberty to deviate from any point of them that may seem at variance with the objects of the expedition, but you are further empowered to take such other steps as shall be desirable at the time, and which are not provided for in these orders.

" Given under our hands, 16th March 1848.

(Signed) " AUCKLAND.

J. W. D. DUNDAS.

" To Sir John Richardson, M. D., &c.

" By command, &c.

(Signed) " W. A. B. Hamilton."

CHAP. II.

THE preceding pages contain an exposition of the
objects of the expedition, with a general outline
of the course to be pursued after leaving the Mackenzie; but as that great river can be attained only
by a long and laborious lake and river navigation,
it is proper that I should introduce the narrative
by a brief account of that first stage of our overland journey.   There are two routes to the Mackenzie, one of which, traced at an early period by
the Canadian fur companies, passes through Lakes
Huron and Superior, the Kamenistikwoya, or Dog
River, the Lake of the Woods, Rainy Lake, Lake
Winipeg, Cedar Lake, the Saskatchewan River,
Beaver and Half-moon Lakes, Churchill or English
River, Isle à la Crosse Buffalo and Methy Lakes
to the Methy Portage, and the Clear-water or Little
Athabasca River, one of the affluents of the Mac-

kenzie. From thence there is a continuous water-course to the sea, through the Elk or Athabasca River, Athabasca Lake, Slave River and Lake, and the Mackenzie proper.

The length of this interior navigation from Montreal to the Arctic Sea is, in round numbers, four thousand four hundred miles, of which sixteen hundred miles are performed on the Mackenzie and its affluents, from Methy Portage northwards, and in which the only interruptions to boat navigation are a few cascades and rapids in Clear-water and Slave Rivers.

During the existence of the North-west, X-Y, and other fur companies trading from Canada, supplies were conveyed to their northern posts by the way of the Ottawa river and great Canada lakes; but they reached the distant establishments on the Mackenzie only in the second summer, having been deposited in the first year at a depôt on Rainy River. Owing to the shallowness of the streams, and badness of the portage roads over the heights between Lake Superior and Rainy Lake, the transport of goods requires to be performed in canoes, with much manual labour, and is, consequently, very expensive. On this account the Hudson's Bay Company, who are now the sole possessors of the northern fur trade, no longer take their trading goods from Canada, but send them by the shorter and cheaper way of Hudson's Bay; though they still employ two or three canoes on the Lake Superior route, to accommodate the Governor in his annual journeys from his residence at La Chine to Norway House, and for the transport of newly-hired servants to the interior, or for bringing down officers coming out on furlough, and men whose period of service has expired. No repairs having of late years been made on the portage roads, they have very much deteriorated, and are truly execrable.

The distance between York Factory in Hudson's Bay and Norway House, situated near the northeast corner of Lake Winipeg, does not much exceed three hundred miles; and as the navigation, though much interrupted by rapids and cascades, admits, in the majority of seasons, of boats carrying a cargo of between fifty and sixty hundred-weight, it offers a much more economical approach to the interior of the fur countries than the other; since one of these boats may be managed by the same crew that is required for a canoe carrying only twenty hundred-weight. The Hudson's Bay ships are generally two in number; one of them being employed in taking supplies to Moose Factory, at the bottom of James's Bay, and the other to York Factory, in latitude 57° N., longitude 92½° W., on the west coast of Hudson's Bay. They sail annually from the Thames on the first Saturday in June, and, after touching at the Orkneys, to receive labourers for the Company's service, proceed on their voyage to Hudson's Straits. The York Factory ship has dropped her anchor at the mouth of Hayes River as early as the 5th of August, and as late as the beginning of September. A tardy arrival is very inconvenient, both in respect of forwarding goods into the interior, and also with regard to the return of the ship to England, there being in such a case scarcely time for the embarkation of the cargo of furs and the passage of Hudson's Straits before the winter sets in.

This brief notice of the modes of communication with Rupert's Land — for so the possessions of the Hudson's Bay Company are named — is given, to explain some parts of the plan of the expedition, and particularly to show why the stores and men were sent out by ships which sailed in June 1847, although the expediency of searching expeditions was not considered by the Admiralty to be established until the last of the whalers came in at the close of that season, without bringing tidings of the discovery ships. It was arranged that in that case, the officers were to leave England early in 1849, and, travelling as rapidly as they could through the United States and Canada, were to overtake the party conveying the stores in the vicinity of Methy Portage.

In April, 1847, I had the advantage of a personal interview with Sir George Simpson, Governor-in-chief of Rupert's Land, who was then on a visit to England, and of concerting with him the measures necessary for the future progress of the expedition; and I may state here that he entered warmly into the projects for the relief of his old acquaintance Sir John Franklin; and from him I received the kindest personal attention, and that support which his thorough knowledge of the resources of the country and his position as Governor enabled him so effectively to bestow. He informed me that the stock of provisions at the various posts in the Hudson's Bay territories was unusually low, through the failure of the bison hunts on the Saskatchewan, and that it would be necessary to carry out pemican from this country, adequate not only to the ulterior purposes of the voyage in the Arctic Sea, but also to the support of the party during the interior navigation in 1847 and 1848. I, therefore, obtained authority from the Admiralty to manufacture, forthwith, the requisite quantity of that kind of food in Clarence Yard; and as I shall have frequent occasion to allude to it in the subsequent narrative, it may be well to describe in this place the mode of its preparation.

The round or buttock of beef of the best quality, having been cut into thin steaks, from which the fat and membranous parts were pared away, was dried in a malt kiln over an oak fire, until its moisture was entirely dissipated, and the fibre of the meat became friable. It was then ground in a malt mill, when it resembled finely grated meat. Being next mixed with nearly an equal weight of melted beef-suet or lard, the preparation of plain pemican was complete; but to render it more agreeable to the unaccustomed palate, a proportion of the best Zante currents was added to part of it, and part was sweetened with sugar. Both these kinds were much approved of in the sequel by the consumers, but more especially that to which the sugar had been added. After the ingredients were well incorporated by stirring, they were transferred to tin canisters, capable of containing 85 lbs. each; and, having been firmly rammed down and allowed to contract further by cooling, the air was completely expelled and excluded by filling the canister to the brim with melted lard, through a small hole left in the end, which was then covered with a piece of tin, and soldered up. Finally, the canister was painted and lettered according to its contents. The total quantity of pemican thus made was 17,424 lbs., at a cost of 1s. 7¼d. a pound. But the expense was somewhat greater than it would otherwise have been from the inexperience of the labourers, who required to be trained, and from the necessity of buying meat in the London market at a rate above the contract price, occasioned by the bullocks slaughtered by the contractor for the naval force at Portsmouth being inadequate to the supply of the required number of rounds. Various temporary expedients were also resorted to in drying part of the meat, the malt kiln and the whole Clarence Yard establishment being at that time fully occupied night and day in preparing flour and biscuit for the relief of the famishing population of Ireland. By the suggestions of Messrs. Davis and Grant, the intelligent chief officers of the Victualling Yard, and their constant personal superintendence, every difficulty was obviated.

As the meat in drying loses more than three fourths of its original weight, the quantity required was considerable, being 35,651 lbs.*; and the sudden abstraction of more than one thousand rounds of beef from Leadenhall Market occasioned speculation among the dealers, and a rise in the price of a penny per pound, with an equally sudden fall when the extra demand was found to be very temporary.†

---

* By drying this was reduced to about 8000 lbs.

† Particulars of the estimated expense of pemican, manu-

The natives dry their venison by exposing the thin slices to the heat of the sun, on a stage, under which a small fire is kept, more for the purpose of driving away the flies by the smoke than for promoting exsiccation; and then they pound it between two stones on a bison hide. In this process the pounded meat is contaminated by a greater or smaller admixture of hair and other impurities. The fat, which is generally the suet of the bison, is added by the traders, who purchase it separately from the natives, and they complete the process by sewing up the pemican in a bag of un-

factured in the Royal Clarence Victualling Yard, in Midsummer quarter, 1847:

				£	s.	d.	£	s.	d.
Fresh beef	35,651 lbs.	at 6¾d. per lb.		979	10	1			
Lard	- 7,549 —	at 88s. per cwt.		296	11	4			
Currants -	1,008 —	at 84s. per cwt.		37	16	0			
Sugar	- 280 —	at 31s.2d. per cwt.		3	17	11			
							1,317	15	4
Oak slab 46 fms. at 22s. 6d. per load				47	5	0			
Hire for labourers	-	-	-	59	8	8			
Hire of kiln and cartage		-	-	8	1	0			
							114	14	8
							1,432	10	0
Deduct for scraps of fat sold	-	-	-	-			35	18	1
							1,396	11	11

Quantity of pemican manufactured 17,424 lbs.; average cost per lb. 1s. 7¼d.

dressed hide with the hairy side outwards. Each of these bags weighs 90 lbs. and obtains from the Canadian voyagers the designation of "un taureau." A superior pemican is produced by mixing finely powdered meat, sifted from impurities, with marrow fat, and the dried fruit of the Amelanchier.

By order of the Admiralty, four boats were built; two of them in Portsmouth Dock Yard, and two in Camper's Yard at Gosport. These boats, to fit them for river navigation, were required to be of as small a draught of water as was consistent with the power of carrying a cargo of at least two tons; to have the head and stern equally sharp, like a whale-boat, that they might be steered with a sweep oar when running rapids; and to be of as light a weight as possible, for more easy transportation across the numerous portages on the route, and especially the formidable one between Methy Lake and Clear-water River. They were also to be as good sea-boats as a compliance with the other requisites would allow. It is manifest that the invention of a form of boat possessing such various and in some respects antagonistic qualities would task the skill of the constructor, and I felt much indebted to William Rice, Esq., Assistant Master Builder of Portsmouth Yard, for the care and skill with which he worked out a successful result. The Company's boats, or barges, as they term them, are generally about 36 feet long from stem to stern-post, 8 feet wide, stoutly framed and planked, and are capable of carrying seventy packages of 90 lbs. each, with a crew of eight men. The thickness of the planks of these boats is such that they sustain with little injury a severe blow against a rock, to which they are much exposed in descending the rapids; but their weight being proportionally great, they are transported with much labour across the ordinary portages, and it is necessary to avoid this operation altogether at Methy Portage by keeping a relay of boats at each terminus. Moreover, these boats resemble the London river barges in the great rake of the stem and stern, by which they are better fitted for the descent of a rapid, but from the flatness of their floors they are leewardly and bad seaboats.

Two of the expedition boats measured 30 feet from the fore part of the stem to the after part of the stern-post, 6 feet in breadth of beam, and 2 feet 10 inches in depth; and each of them weighed $6\frac{1}{2}$ cwt., or, including fittings, masts, sails, oars, boat-hook, anchor, lockers and tools, half a ton. The other two boats measured 28 feet in length, 5 feet 6 inches in width, 2 feet 8 inches in depth; and weighed $5\frac{1}{4}$ cwt., or, with the moveable fittings and equipment, 9 cwt. They were all clinker-built of well-seasoned Norway fir planks $\frac{6}{16}$ of an inch thick; ashen floors placed 9 inches apart; stem, stern-posts, and knees of English oak; and gunwales of rock-elm. To admit of their stowing the requisite cargo, they were necessarily very flat-floored, but screws and bolts were fitted to the kelson, by which a false keel might be readily bolted on before they reached the Arctic Sea, so as to render them more weatherly. The larger boats when quite empty drew $7\frac{1}{4}$ inches of water, and, when loaded with two tons but without a crew, $14\frac{1}{4}$ inches. They were constructed of two sizes, that the smaller might stow within the larger ones during the passage across the Atlantic.

For the voyage on the Arctic Sea, a crew of five men to each boat was considered sufficient, but for river navigation a bowman and steersman experienced in the art of running rapids were required in addition. Five seamen and fifteen sappers and miners were selected in the month of May, for the expedition, from a number of volunteers. They were all men of good physical powers, and, with one exception, bore excellent characters in their respective services. The solitary exception was one of the sappers and miners who had repeatedly appeared on the defaulters' list for drunkenness, but as he was reported to be in other respects a good and willing workman, and I knew that he would have no means of obtaining intoxicating drinks in Rupert's Land, I yielded to his request that I would allow him an opportunity of retrieving his character. Few seamen were employed, since I knew from experience that as a class they march badly, particularly when carrying a load, and the bulk of the party was composed of sappers and miners, because that corps contains a large proportion of intelligent artizans. Of the men selected, six were joiners or sawyers, and four were blacksmiths, armourers, or engineers, who could be useful for repairing the boats, working up iron, constructing the buildings of our winter residence, or making the furniture.

Every thing was ready before the appointed day; and the boats and stores, having been sent round from Portsmouth to the Thames, were embarked with the expedition men on board the "Prince of Wales" and "Westminster," bound to York Factory, the exigences of the Hudson's Bay trade of that year requiring two ships to go to that port. The stores consisted of 198 canisters of pemican, each weighing 85 lbs., 10 bags of flour, amounting in all to 8 cwt., 5 bags of sugar, weighing 4½ cwt., 2 of tea, weighing 88 lbs., 3 of chocolate, weighing 2 cwt., 10 sides of bacon, amounting to 4½ cwt., and 6 cwt. of biscuit; also 400 rounds of ball cartridge, 90 lbs. of small shot, and 120 lbs. of fine powder in 4 boat magazines. In the arm-chests and lockers of the boats, there were stowed a musket fitted with a percussion lock for each man, with a serrated bay- onet that could be used as a saw; also a complete double set of tools for making or repairing a boat, a tent for each boat's crew, towing-lines, anchors, and one seine net.

Each man was provided with a Flushing jacket and trowsers, a stout blue Guernsey frock, a waterproof over-coat, and a pair of leggins. In- structions were also given that they should be furnished in winter with such moccasins and leather coats as the nature of their employment should render necessary. Could the expedition have depended on procuring supplies of provision at the Company's posts during their progress through the interior, and a sufficient quantity of pemican at one of the northern depôts for the sea voyage, the boats would have been lightly laden, and a quick advance into the interior might have been anticipated. But such not being the case, it was necessary to employ one of the Company's barges to assist in the transport; and Governor Sir George Simpson undertook to provide one, and to engage a proper crew in Rupert's Land, together with bowmen and steersmen for the expedition boats. He also agreed to select from the Company's stores a complete assortment of nets and other necessaries for the use of the party in the winter of 1847-8.*

The Company's ships sailed from the Thames on the 15th of June, 1847, and, being much delayed by ice in Hudson's Straits, had a long passage; so that the "Prince of Wales" did not cross the bar of Hayes River till the 25th of August, nor the "Westminster" until five days later; and the 8th of September arrived before the expedition stores were landed. Sir George Simpson, on his annual visit to the Company's depôt at Norway House, had engaged a guide or river pilot, with the requisite number of bowmen, steersmen, and fishermen, and placed the whole under the superintendence of Mr. John Bell, chief trader, who, having resided many years on the Mackenzie, was intimately acquainted with the natives inhabiting that part of the country. Notwithstanding the high wages offered, being much in advance of the rate ordinarily paid by the Company, and though none of these men were required to extend their services beyond the winter quarters of the party in 1848, there was a scarcity of volunteers; and several of the steersmen, that were, from the necessity of the case, engaged, were men of little experience. None of them were acquainted with the neighbourhood of Great Bear Lake, and they all anticipated with more

* See Appendix.

or less apprehension a season of extreme hardship in that northern region.   Mr. Bell's party consisted of twenty Europeans, a guide, and sixteen Company's voyagers, together with the wives* of three of the latter, and two children; making in all, with himself and two of his own children, forty-five individuals, embarked in five boats.   Had the ships arrived early, there was a possibility of the party reaching Isle à la Crosse before the navigation closed, which, in that district, may be expected to occur about the 20th of October.   But the very late date at which the stores were disembarked precluded such a hope; and the extreme dryness of the season, and consequent lowness of the rivers between York Factory and Lake Winipeg, obliged Mr. Bell to leave a quantity of the pemican and some other packages at York Factory, that he might reduce the draught of his boats.

These facts were communicated to me on the return of the Hudson's Bay ships to England in October; and in February, 1848, I heard by letters forwarded through Canada, that Mr. Bell and his party had, from the causes specified, made slow progress; that the boats had been often

---

* It is desirable to have two or three females at every post in the interior for washing, making, and mending the people's clothes and mocassins, netting snow shoes, making and repairing fishing nets, and other services of a similar nature.

stranded and broken in the shallow waters, causing frequent detention for repairs; and that the party was overtaken by winter in Cedar Lake. Mr. Bell forthwith housed the boats, constructed a store-house for the goods, left several men to take care of them, and such of the women and children as were unable to travel over the snow. This being done, he set out with the bulk of the party for Cumberland House, and reached it on the eighth day after leaving Cedar Lake. His first care was to establish a fishery, which he did on Beaver Lake, two days' walk further north; and having sent a division of the men thither, the others were distributed to the several winter employments of cutting firewood, driving sledges with meat or fish, and such-like occupations. The unforeseen stoppage of the boats occasioned a large consumption of the pemican destined for the sea voyage, but was attended by no other bad consequences, and the deficiency was amply made up in spring through the exertions of the gentlemen in charge of the Company's provision posts on the Saskatchewan; so that Mr. Bell, when he resumed his voyage northwards in the summer of 1848, was enabled to take with him as much of that kind of food as his boats could stow.

While the body of the party was thus passing the winter at Cumberland House and its vicinity,

I was almost daily receiving letters from officers of various ranks in the army and navy, and from civilians of different stations in life, expressing an ardent desire for employment in the expedition. It may interest the reader to know that among the applicants, there were two clergymen, one justice of peace for a Welsh county, several country gentlemen, and some scientific foreigners, all evidently imbued with a generous love of enterprise, and a humane desire to be the means of carrying relief to a large body of their fellow creatures. But as long as there remained a hope of the return of the discovery ships in the autumn of 1847, it was not thought necessary to take any steps for the appointment of a second officer to the party which I was to command. In November, however, when the last whalers from Davis's Straits had come in, I suggested to the late Lord Auckland, then the First Lord of the Admiralty, that Mr. John Rae, chief trader of the Hudson's Bay Company, was fully qualified for the peculiar nature of the service on which we were to be employed. He had resided upwards of fifteen years in Prince Rupert's Land, was thoroughly versed in all the methods of developing and turning to advantage the natural products of the country, a skilful hunter, expert in expedients for tempering the severity of the climate, an accurate observer with the sextant and

other instruments usually employed to determine
the latitude and longitude, or the variations and
dip of the magnetic needle, and had just brought to
a successful conclusion, under circumstances of very
unusual privation, an expedition of discovery fitted
out by the Hudson's Bay Company, for the purpose
of exploring the limits of Regent's Inlet.   Lord
Auckland highly approved of my suggestion, and
Mr. Rae was appointed with the assent of the
Governor and Committee of the Hudson's Bay Company.

Mr. Rae and I left Liverpool on the 25th of
March, 1848, in the North American mail steam-
packet " Hibernia," and landed at New York on the
morning of the 10th of April.   In addition to our
personal baggage, we took with us a few very portable astronomical instruments required for determining our positions; and four pocket chronometers,
one of them being the property of Mr. Frodsham,
which had been used on the several expeditions of
Sir W. E. Parry and Sir John Ross, and which he
wished to lend gratuitously for service in the present enterprise.   We had also a few meteorological
instruments, and some others for determining questions in magnetism, that shall be more particularly
described hereafter, when their employment comes
to be mentioned.   An ample supply of paper for

botanical purposes, a quantity of stationery, a small selection of books, a medicine chest, a canteen, a compendious cooking apparatus, and a few tins of pemican, completed our baggage, which weighed in the aggregate, above 4000 lbs.

Mr. Barclay, the British consul, assisted with much kindness in expediting our departure from New York. An order from the United States Treasury directed that our baggage should not be inspected by the custom-house agents, and it was without delay consigned to the care of Messrs. Wells and Co., forwarders, who contracted to send it to Buffalo, by rail-road, and from thence to Detroit and Saut Sainte Marie, by the first steam-boat, which was advertised to sail from Detroit on the 21st of April. Immediately on landing, the chronometers were placed in the hands of Mr. Blount, of Water Street, that he might ascertain their rate by comparison with the astronomical clock in the observatory. For this service Mr. Blount would receive no remuneration, but, on the contrary, said that he was glad of the opportunity it afforded him of showing his sense of the courtesy he had experienced from the hydrographer of the British Admiralty.

We received the chronometers next day, and embarked in the evening on board the "Empire," for Albany and Troy, with the view of proceeding, by way of Lake Champlain, to Montreal, where the canoe-men engaged for us by Sir George Simpson were ordered to rendezvous.

We waited one day at Whitehall, for the complete disruption of the ice on Lake Champlain *, and did not reach Montreal till the fourth day after leaving New York. Sir George Simpson received us, with his usual kindness and hospitality, at his residence in La Chine, and expedited our arrangements by all the means in his power; but two days were spent in collecting the voyagers † who were engaged as our canoe-men. Four of them, with the levity of their class, were absent at the time finally fixed for our departure, thereby, in terms of their agreements, incurring fines, which were afterwards levied by the Hudson's Bay Company.

The steamers commenced running on the St. Lawrence on the 18th of April; we embarked on the 19th, reached Buffalo on 21st, Detroit on the 23rd, and Saut Ste. Marie, at the outlet of Lake Superior, on the 29th, where we again found our-

---

* The ice broke up on Lake Champlain on the 13th of April. On the previous day a steamer was prevented from reaching Whitehall by drift ice filling a narrow passage of the lake.

† The Canadian term " voyageurs " is usually employed to designate these men, as that is the language in which they are addressed ; but there seems to be no reason why they should not be called " voyagers," or " canoe-men," in an English work.

selves in advance of the season, the Lake being covered with drift ice.*   -

At the Hudson's Bay House, the residence of Chief Factor Ballenden, we found two "north canoes," made ready for us, by direction of Sir George Simpson, and, having engaged four additional men to supply the place of an equal number who had failed to appear at La Chine, our crews now consisted of

First Canoe.

Thomas Karahonton (*dit* Gros Thomas), an Iroquois guide.				
Laxard Tacanajazè	-	-	-	Iroquois.
Thomas Nahanajazè	-	-	-	,,
François Monegon	-	-	-	,,
Thomas Anackera	-	-	-	,,
Sauveur St. Martin	-	-	-	Canadian.
Thomas Cadrant	-	-	-	Half-breed.
Joseph Dinduvant	-	-	-	,,

Second Canoe.

Charlot Arahota	-	-	-	Iroquois.
Louis Taranta	-	-	-	,,
Ignace Atawackon	-	-	-	,,
Ignace Sataskatchi	-	-	-	,,
Apoquash	-	-	-	Chippeway.
Miskiash	-	-	-	,,
Piquatchiash (Peter)	-	-	-	,,

---

* In the instructions the route by Penetanguishene is specified for the expedition to take; but the steamer from that port to Saut Ste. Marie was not advertised to start for three weeks later than our time.

Two days were occupied in re-packing our bag-gage, instruments, and provisions, in cases weigh-ing 90 lbs. each (being the established size for the portages); in which, and in all other matters connected with our equipment and comfort, we experienced great assistance and personal kindness from Mr. Ballenden.   On the 2nd of May, 1851, we quitted his hospitable roof, but it was the 4th before the ice on the lake broke up, and permitted us to pass the portal of the lake formed by Gros Cap and Point Iroquois.

We accomplished the navigation of the lake on the 12th by arriving at Fort William, attained the summit of the water-shed which separates the St. Lawrence and Winipeg valleys on the 18th *, the mouth of the River Winipeg on the 29th, Norway House, near the efflux of Nelson River, on the 5th of June, and Cumberland House, on the Saskat-chewan, on the 13th; our passage through Lake Winipeg having been much delayed by ice, from which we did not disengage ourselves till the 9th.

We learnt at Cumberland House, that Mr. Bell had given the boats a thorough repair at Cedar Lake in the spring, had brought them and the

* Dog Lake, near the summit of this water-shed, broke up only on the eve of our arrival; an Indian whom we met on the Kamenistikwoya, which flows from it, having crossed it on the preceding day over the ice.

stores up on the first opening of the Saskatchewan, and was now a fortnight in advance of us on his way to Methy Portage.   The bulk of his party had been maintained at Beaver Lake on fish, but some having wintered in Cedar Lake, to look after the stores, and the fishery there having failed, there had been an unavoidable consumption of the pemican destined for the sea-voyage.   The provision posts on the upper part of the Saskatchewan had fortunately been able to replace what was consumed, and Mr. Bell had started from Cumberland House with his boats fully laden.

He had left two men of the English party behind, who were unequal to the labours of the voyage; one of them, because of an injury received in the hand early in the spring, and the other owing to a recurrence of pains in the bones, with which he had formerly been afflicted.   After carefully examining these men, I decided upon sending them to York Factory by the first conveyance which offered, that they might return to England in September, in the Hudson's Bay annual ship.

Having thus briefly touched on the line of route pursued by us in a journey of two thousand eight hundred and eighty statute miles, from New York to the wintering place of the boat-party *, I shall

---

* New York to La Chine   -   -   - 500 miles.
  La Chine to Buffalo   -   -   - 372   „

detail the events of the remainder of the voyage in form of a daily journal. To have given a full account of the country travelled through between New York and the Saskatchewan, would have swelled the work to an inconvenient size; and I must, therefore, refer the reader, who wishes to have a physical description of that part of the continent, to Sir Charles Lyell's accounts of his recent visits to the United States, to Professor Agassiz's description of Lake Superior, and to Major Long's voyage to the St. Peter's, Red River, and River Winipeg. The Appendix to the present work also contains a summary of the physical geography of North America, wherein the lake basins of the St. Lawrence and Winipeg or Saskatchewan are particularly noticed. This may be consulted by the reader before he enters upon the narrative of the voyage, and I shall give in this place a few remarks, by way of preface to the botanical and geological notices which follow in the journal.

On the bluff granitic promontories and bold acclivities which form the northern shore of Lake

Buffalo to Detroit	-	-	-	230 miles
Detroit to Saut Ste. Marie	-	-	400	,,
Saut Ste. Marie to Fort William	-	-	370	,,
Fort William to Cumberland House (Franklin's second journey)	-	-	-	1,018 ,,
				2,880 ,,

Superior, the forest is composed of the white spruce, balsam fir, Weymouth pine, American larch, and canoe birch, with, near the edge of the lake and on the banks of streams, that pleasant inter-mixture of mountain maple and dogwood* which imparts such a varied and rich gradation of orange and red tints to the autumnal landscape. Other trees exist, but not in sufficient numbers to give a character to the scenery. Oaks are scarce, and beech disappears to the south of the lake. The American yew, which does not rise into a tree like its European namesake, is the common underwood of the more fertile spots, where it grows under the shade to the height of three or four feet, in slender bush-like twigs. On the low sandstone islands deci-duous trees, such as the poplars and maples, abound, with the nine-bark spiræa, cockspur thorns, wil-lows, plums, cherries, and mountain-ash.† When we entered the lake on the 4th of May, large accu-mulations of drift snow on the beaches showed the lateness of the season; none of the deciduous trees had as yet budded; and the precocious catkins of a silvery willow (*Salix candida*), with the humble

* *Abies alba, Abies balsamea, Pinus strobus, Larix ameri-cana, Betula papyracea, Acer montanum,* and *Cornus alba.*

† *Populus tremuloides* et *balsamifera ; Acer ; Spirea opuli-folia ; Cratægus crus-galli, punctata, glandulosa,* et *coccinea ; Prunus americana ; Cerasus pumila, nigra, pennsylvanica, virginiana,* et *serotina ; Pyrus americana.*

flowers of a few Saxifrages and Uvulariæ, gave the
only promises of spring.

In various parts of the lake, the gorges lying between the jutting bluffs of granite or slate are filled
with deposits of sand rising in four or five successive terraces to the height of more than a hundred
feet above the present surface of the water.   Mr.
Logan has measured some of the most remarkable,
and Professor Agassiz devotes an interesting chapter
to the discussion of their origin ; in which he comes
to the conclusion that they were formed by the
waters of the lake itself, and have been raised, at
various intervals, from the beach to their present
levels, by the agency of the innumerable trap dikes,
which cross the rocks in many directions.

Near Cape Choyyè, on the south side of Michipicoten Bay, a small gorge between two points of
granite is filled, to the height of twenty-five feet
above the water, with rolled stones and pebbles.
These rounded stones vary in size from that of a
hogshead to a hen's egg, and form a steeply shelving beach, with a flat terraced summit, the larger
boulders being next the water, and the smaller
pebbles highest up.   As the cove is sheltered from
high waves, the terrace could not be thrown up by
the waters of the lake standing at their present
height ; nor can it be owing to the pressure of ice,
since that would not graduate the pebbles.

At Michipicoten River we had a curious illustration of the agency of frost, on the outlet of the stream.   During the summer, when the waters are low, the waves of the lake throw a sandy bar across the mouth of the river.   In winter this bar freezes into a solid rock and closes the channel, but as the spring advances the stream acts upon it and cuts a passage.   At the time of our visit, on May 7th, the river was in flood, and the bar remained hard, but was cleft by a narrow channel with precipitous sides like sandstone cliffs, and a cascade one foot high existed.   This fall, which was five or six feet high when the river broke, would, we were told, entirely disappear in a few days.

The north coast of Michipicoten Bay is the boldest and most rugged of the shores of the lake, and apparently the least capable of cultivation. It rises to the height of about eight hundred feet, and for twenty-five miles comes so precipitously down to the water that there is no safe landing for a boat.   On much of the crags the forest was destroyed by fire, many years ago, and with it the soil, presenting a scene of desolation and barrenness not exceeded on the frozen confines of the Arctic Sea.   The few dwarf trees that cling to the crevices of the rocks, struggling, as it were, between life and death, add to the dreariness of the prospect rather than relieve it, and wreaths of drift snow lining many of the recesses, at the time

when we passed, though it was in the second week of
the glorious month of May, gave a most unfavourable impression of the land and its climate.   Professor Agassiz has pointed out the sub-arctic character
of the vegetation of Lake Superior, by a lengthened
comparison with the subalpine tracts of Switzer-
land ; but this is due to the nature of the soil, rather
than to the elevation or northern position of the district ; for as we advance to the north at an equal
elevation above the sea, but more to the westward,
so as to enter on silurian or newer deposits in the
vicinity of the Lake of the Woods and Rainy River,
we find *cacti* and forests having a more southern
aspect.

The ascent to the summit of the water-shed between Lakes Superior and Winipeg, by the Kamenistikwoya River, is made by about forty portages,
in which the whole or part of a canoe's lading is
carried on the men's shoulders ; and a greater
number occur in the descent to the Winipeg.   The
summit of the water-shed is an uneven swampy granitic country, so much intersected in every direction by lakes that the water surface considerably
exceeds that of the dry land.   Its mean elevation
above Lake Superior is about eight hundred feet,
and the granite knolls and sand-banks, which vary
its surface, do not rise more than one hundred and
fifty or two hundred feet beyond that general level,
though their altitude above the river valleys which

surround them is occasionally greater, giving the
district a hilly aspect. The highest of these
eminences does not overtop Thunder Mountain and
some other basalt-capped promontories on Lake
Superior, and had not the silurian strata, which,
judging by the patches which remain, once covered
the gneiss and granitic rocks nearly to their summits,
been removed, the country would have been almost
level, and would have formed part of the rolling
eastern slope of the continent, above whose plane
the highest of the hills on Lake Superior scarcely
rises. The summit of this water-shed of the St.
Lawrence basin, commencing towards the Labrador
coast, runs south 52° west, or about south-west half-
west, at the distance of rather more than two
hundred miles from the water-course, until it comes
opposite to that elbow of the line of the great lakes
which Lake Erie forms; it then takes a north 51°
west course, or about north-west half-west, towards
the north-east end of Lake Winipeg, and onwards
from thence in the same direction to Coronation
Gulf of the Arctic Sea. The angle at which the two
arms of this extensive water-shed (but no where
mountain ridge) meet between Lakes Huron and
Ontario is within half a point of a right one, and
the character of the surface is everywhere the same,
bearing, in the ramifications and conjunctions of its
narrow valleys filled with water, no distant resem-

blance to the fiords of the Norway coast. Such a preponderance of fresh water, coupled with the tardy melting of the ice in spring, makes a late summer, and augments the severity of the climate beyond that which is due to the northern position of the district. Though the whole tract is most unfavourable for agriculture, much of the scenery abounds in picturesque beauty. Of this we have an instance in the Thousand Islands Lake, which forms the funnel-shaped outlet of Lake Ontario. At this place the pyrogenous rocks, denuded of newer deposits, cross the river to form a junction with the lofty highlands of the northern counties of New York. The round-backed, wooded hummocks of granite which constitute the more than thousand islets of this expanse of water, are grouped into long vistas, which are alternately disclosed and shut in as we glide smoothly and rapidly among them, in one of the powerful steamers, that carry on the passenger traffic of the lakes. The inferior fertility of this granite belt has deferred the sweeping operations of the settler's axe; the few farm-steadings scattered along the shore enhance the beauty of the forest; and the eye of the traveller finds a pleasant relief in contemplating the scenery, after having dwelt on the monotonous succession of treeless clearances lower down the river. Sooner or later, however, the shores of the Lake of the

Thousand Isles will be studded with the summer retreats of the wealthy citizens of the adjacent states, and the incongruities of taste will mar the fair face of nature.

On the summit of the canoe-route between Lakes Superior and Winipeg, a sheet of water, bearing the analogous appellation of Thousand Lakes, is also studded with knolls of granite, forming islets; but low mural precipices are more common there; and there is, moreover, an inter-mixture of accumulations of sand, such as are commonly found on the summit of the water-shed, along its whole range. The general scenery of this lake is similar to that of the Thousand Islands; but though the elevation above the sea does not exceed fourteen hundred feet, the voyagers say that frosts occur on its shores almost every morning throughout the summer.

Silurian strata occur on both flanks of both arms of the water-shed above spoken of, to a greater or smaller extent throughout their whole length.* When we descend to Lake Winipeg we come upon epidotic slates, conglomerates,

---

* A *Pentamerus*, very like *P. Knightii*, was gathered by Dr. Bigsby on the Lake of the Woods, and presented by him to the British Museum. He probably found it in some of the western arms of the lake, the islands in the more easterly part being mostly granite.

sandstones, and trap rocks, similar to those which occur on the northern acclivity of the Lake Superior basin; and after passing the straits of Lake Winipeg, we have the granite rocks on the east shore, and silurian rocks (chiefly bird's-eye limestone) on the west and north, the basin of the lake being mostly excavated in the limestone. The two formations approach nearest to each other at the straits in question, where the limestone, sandstone, epidotic slates, green quartz-rock, greenstone, gneiss, and granite, occur in the close neighbourhood of each other.

The eastern coast-line of Lake Winipeg is in general swampy, with granite knolls rising through the soil, but not to such a height as to render the scenery hilly. The pine forest skirts the shore at the distance of two or three miles, covering gently-rising lands, and the breadth of continuous lake-surface seems to be in process of diminution, in the following way. A bank of sand is first drifted up, in the line of a chain of rocks which may happen to lie across the mouth of an inlet or deep bay. Carices, balsam-poplars, and willows, speedily take root therein, and the basin which lies behind, cut off from the parent lake, is gradually converted into a marsh by the luxuriant growth of aquatic plants. The sweet gale next appears on its borders, and drift-wood, much of it rotten and comminuted, is thrown up on the exterior bank, together with some roots and stems of larger trees. The first spring storm covers these with sand, and in a few weeks the vigorous vegetation of a short but active summer binds the whole together by a network of the roots of bents and willows. Quantities of drift-sand pass before the high winds into the swamp behind, and, weighing down the flags and willow branches, prepare a fit soil for succeeding crops. During the winter of this climate, all remains fixed as the summer left it; and as the next season is far advanced before the bank thaws, little of it washes back into the water, but, on the contrary, every gale blowing from the lake brings a fresh supply of sand from the shoals which are continually forming along the shore. The floods raised by melting snows cut narrow channels through the frozen beach, by which the ponds behind are drained of their superfluous waters. As the soil gradually acquires depth, the balsam-poplars and aspens overpower the willows, which, however, continue to form a line of demarcation between the lake and the encroaching forest.

Considerable sheets of water are also cut off on the north-west side of the lake, where the bird's-eye limestone forms the whole of the coast. Very recently this corner was deeply indented by narrow, branching bays, whose outer points were limestone cliffs. Under the action of frost, the thin horizontal beds of this stone split up, crevices are formed perpendicularly, large blocks are detached, and the cliff is rapidly overthrown, soon becoming masked by its own ruins. In a season or two the slabs break into small fragments, which are tossed up by the waves across the neck of the bay into the form of narrow ridge-like beaches, from twenty to thirty feet high. Mud and vegetable matter gradually fill up the pieces of water thus secluded; a willow swamp is formed; and when the ground is somewhat consolidated, the willows are replaced by a grove of aspens.* Near the First and Second Rocky Points†, the various stages of this process may be inspected, from the rich alluvial flat covered with trees and bounded by cliffs that once overhung the water, to the pond recently cut off by a naked barrier of limestone, pebbles, and slabs, discharging its spring floods into the lake, by a narrow though rapid stream. In some exposed places the pressure of the ice, or power of the waves in heavy gales, has forced the

* The fact of the formation of these detached ponds, marshes, and alluvial flats points either to a gradual elevation of the district, or to an enlargement of the outlet of the lake, producing a subsidence of its waters.

† The strata at these points contain many gigantic orthoceratites, some of which have been described by Mr. Stokes in the Geological Transactions.

VOL. I.   F

limestone fragments into the woods, and heaped them round the stems of trees, some of which are dying a lingering death; while others, that have been dead for many years, testify to their former vitality, and the mode in which they have perished, by their upright stems, crowned by the decorticated and lichen-covered branches which protrude from the stony bank. The analogy between the entombment of living trees, in their erect position, to the stems of *sigillariæ*, which rise through different layers in the coal-measures, is obvious.*

The action of the ice in pushing boulders into the woods was observed at an earlier period of our voyage, and is noticed in the following terms in my journal. "In the first part of our course through Rainy Lake we followed a rocky channel, which was in many places shallow, and varied in breadth from a mile, down to a few yards. Some long arms stretch out to the right and left of the route, and particularly one to the eastward, into which a fork of Sturgeon River is said to enter. There is considerable current in these narrows.

* If one of the spruce firs included in the limestone debris had its top broken off, and a layer of mud were deposited over all, we should have the counterpart of a sketch in Sir Henry de la Beche's Manual (p. 407.). The thick and fleshy rhizomata of the *Calla palustris*, marked with the cicatrices of fallen leaves, and which are abundant in these waters, bear no very distant resemblance to *stigmariæ*.

The first expanse of water we traversed is six miles across, and the second is fully wider. They are connected by a rocky channel, on whose shores many boulders are curiously piled up eight or ten feet above the rocks on which they rest. Other boulders lie in lines among the trees near the shore. They have been thrust up, many of them very recently, by the pressure of the ice, since the channel is too narrow for the wind to raise waves powerful enough to move such stones."

The granite and gneiss which form the east shore of Lake Winipeg strike off at its north-east corner, and, passing to the north of Moose Lake, go on to Beaver Lake, where the canoe-route again touches upon them. At some distance to the westward of them the Saskatchewan, which is the principal feeder of Lake Winipeg, flows through a flat limestone country, which is full of lakes, the reticulating branches of the river, and mud-banks: it has in fact all the characters of a delta, though the divisions of the stream unite into one channel before entering the lake. This flat district extends nearly to the forks of the river, above which the prairie lands commence. Pine Island Lake, Muddy Lake, Cross Lake, and Cedar Lake, where the boats were arrested by ice in 1848, are dilatations of the Saskatchewan, and when the water rises a very few feet, the whole district is flooded;

which commonly occurs on the snow melting in
spring.   Some way to the south lies an eminence
of considerable height, named by the Crees *Wapŭs-
këow-watchi* *, and by the Canadians *Basquiau.*
It separates Winepegoos Lake, and Red-Deer Lake
and River from the bed of the Saskatchewan.   I
am ignorant of its geological structure, not having
visited it.

With respect to the forests : The white or sweet
cedar (*Cupressus thyoides*) disappears on the south
side of Rainy Lake, within the American boundary line.   The Weymouth pine, various maples,
cockspur thorns, and the fern-leaved Comptonia,
reach the southern slope of the Winipeg basin.
Oaks extend to the islands and narrows of that lake.
The elm, ash, arbor vitæ, and ash-leaved maple terminate on the banks of the Saskatchewan.   The
" wild rice," or *Folle avoine*† of the voyagers and
traders, grows abundantly in the district between
Lakes Superior and Winipeg.   This grain resembles
rice in its qualities, but has a sweeter taste.   Though
small, it swells much in cooking, and is nourishing,
but its black husk renders it uninviting in its
natural state.   In favourable seasons it affords
sustenance to a populous tribe of Indians, but the

---

* *Wapŭs*, strait ; *Ke-ow*, woods ; *Watchi*, hill : the signification being, " a pass through woods on a hill."

† *Zizania aquatica* L., or *Hydropyrum esculentum* of Link.

supply is uncertain, depending greatly on the height of the waters. In harvest time the natives row their canoes among the grass, and, bending its ears over the gunwale, thresh out the grain, which separates readily. They then lay it by for use in neatly-woven rush baskets. This grass finds its northern limit on Lake Winipeg, and it is common in the western waters of the more northern of the United States; but how far south it extends, I have not been able to learn. Strachey, in his " Historie of Travaille in Virginia," speaks of a " graine called *Nattowine*, which groweth as bents do in meadowes. The seeds are not much unlike rice, though much smaller; these they use for a deyntie bread, buttered with deere's suet." (p. 118.) It is possible that he may refer to a smaller species (*H. fluitans*) of the same genus, which is known to abound in Georgia; but the seed of that could scarcely be collected in sufficient quantity. The hop plant (*Humulus lupulus*) reaches the south end of Lake Winipeg, and, according to Mr. Simpson, yields flowers plentifully in the Red River colony. We observed it in the autumn of 1849 growing luxuriantly on the banks of the Kamenistikwoya, and connecting the lower branches of the trees with elegant festoons of fragrant flowers. An opinion prevailed among the traders that Lord Selkirk introduced it into this neighbourhood when he took possession of the North-west Company's post of Fort William, upwards of thirty years ago; but the plant is indigenous to America, and grows abundantly in the Raton Pass, lying on the 37th parallel, at the height of eight thousand feet above the sea, as well as in many localities of the northern States. Throughout the canoe-route from Lake Superior to Lake Winipeg, no district shows such fertility as the banks of Rainy River. In autumn, especially, the various maples, oaks, sumachs, ampelopsis, cornel bushes, and other trees and shrubs whose leaves before they fall assume glowing tints of orange and red, render the woodland views equal, if not superior, to the finest that I have seen elsewhere on the American continent, from Florida northwards. Nor are showy asters, *helianthi, lophanthi, gentianeæ, physostegiæ, irides,* and many other gay flowers, wanting to complete the adornment of its banks.

From Saut Ste. Marie to the Saskatchewan, and the banks of Churchill River, the native inhabitants term themselves *In-ninyu-wuk* or *Ey-thinyu-wuk,* and are members of a nation which formerly extended southwards to the Delaware. That part of this widely spread people which occupies the north side of Lake Huron, the whole border of Lake Superior, and the country between it and the south end of Lake Winipeg, call themselves

Ochipewa, written also *Ojibbeway*, or Chippeway*;
and the more northerly division, who name themselves *Nathè-wywithin-yu*, are the Crees of the
traders, and Knistenaux of French writers. In
a subsequent chapter I shall speak more particularly of the place which this people hold among
the aboriginal nations. At present, I wish merely
to point out some of the circumstances which have
tended to work out a difference in the moral character of these two tribes, essentially the same
people in language and manners. The Crees have
now for more than twenty-six years been under
the undivided control and paternal government
of the Hudson's Bay Company, and are wholly
dependent on them for ammunition, European
clothing, and other things which have become
necessaries. No spirituous liquors are distributed
to them, and schoolmasters and missionaries are
encouraged and aided by the Company, to introduce among them the elements of religion and
civilization. One village has been established near
the depôt at Norway House, and another at the
Pas on the Saskatchewan, each having a church,
and school-house, and a considerable space of cultivated ground. The conduct of the people is
quiet and inoffensive ; war is unknown in the Cree

* They are the *Sauteurs* or *Saulteaux* of the Canadians, and
Sotoos of the fur traders.

district; and the Company's officers find little difficulty in hiring the young men as occasional
labourers.

The case is otherwise with the Chippeways, who
live within the Company's territories. The vicinity of the rival United States Fur Company's
establishments; the vigorous competition which
is carried on between them and the Hudson's
Bay Company, in prosecution of which spirituous
liquors are dispensed by both parties liberally to
the natives; and the abundance of *Folle avoine* on
Rainy River and the River Winipeg, with the
plentiful supply of sturgeon obtained from these
waters, rendering the natives independent of either
party, have had a demoralising effect, and neither
Protestant nor Roman Catholic missionaries have
been able to make any impression upon them.
One party of these Indians, from whom we purchased a supply of sturgeon on Rainy River, are
briefly characterised in my notes, made on the
spot, as being " fat, saucy, dirty, and odorous."
A Roman Catholic church, erected some years ago
on the banks of the Winipeg, has been abandoned,
with the clearing around it, on account of the want
of success of the priest in his endeavours to convert
the natives; and neither the Hudson's Bay Company nor the United States people have been able
to extinguish the deadly feud existing between the

Chippeways and Sioux, nor to restrain their war parties.

Very recently the Chippeways of Lake Superior, through some oversight in the Canadian government in not making arrangements with them at the proper time, organised a war party against the mining village of Mica Bay, containing more than a hundred male inhabitants. In passing through Lake Superior we were pleased with the flourishing appearance of this village, containing many nicely white-washed houses, grouped in terraces on the steep bank of the lake. The mines were worked by a company, under a grant from the Canadian legislature, who, at the same period, made many other similar grants of mining localities on the lake, without previously purchasing the Indian rights. As the game is nearly extinct on the borders of the lake, the natives subsist chiefly by the fisheries; and the vicinity of the mining establishments was likely to be beneficial to them rather than injurious, by providing a market for their fish. But when they beheld party after party of white men crowding to their lands, eager to take possession of their lots by erecting buildings, and inquisitively examining every cliff, they acquired exaggerated ideas of the value of their rocks. For two summers they descended in large bodies to Saut Ste. Marie, expecting payment, and, being disappointed, thought they were trifled with. They determined, therefore, in council, to bring matters to a crisis by expelling the aggressors, and, in the autumn of 1849, made a descent upon Mica Bay, and drove away the miners and their families. To repel this attack a regiment was ordered up from Canada, at an expense which would have paid the Indians again and again: but a small part of the force only reached Mica Bay, to find the Chippeways gone; the rest were driven back to Saut Ste. Marie by stormy weather, not without very severe suffering, leading, I have been informed, to loss of life.

CHAP. III.

WE left Cumberland House at 4 A. M., on the 14th of June, but had not passed above three miles through Pine Island Lake, before we were compelled to seek shelter on a small island by a violent thunder storm, bringing with it torrents of rain. The rain moderating after a few hours, we resumed our voyage; but the high wind continuing and raising the waves, our progress was slow, and the day's voyage did not exceed twenty-two miles. In the part of the lake where we encamped the limestone (silurian*) rises, in successive outcrops,

* Some fragments of large *Orthocerata*, and a specimen of *Receptaculites neptunii*, point to the bird's-eye and Trenton

to the height of thirty feet above the water, the
strike of the beds being about south-west by west,
and north-east by east, or at right angles to the
general line of direction of the gneiss and granite
formation, which lies to the eastward. Many
boulders of granite and trap rocks are scattered
over the surface of the ground, far beyond the reach
of any modern means of transport.

Thunder and heavy rain detained us in our
encampment the whole of the following day; but
some improvement in the weather taking place at
midnight, we embarked, and at one in the morning
of the 16th entered Sturgeon River, named by the
voyagers, on account of its many bad rapids, " *La
Rivière Maligne.*" We made two portages, and an
hour after noon reached Beaver Lake. The entire

limestones as occurring in this neighbourhood. Mr. Woodward
says of the latter specimen, " The only wood-cut in the New
York State Surveys at all resembling your engine-turned fossil,
is a very rude representation of part of a circular disk, with
radiating and *concentric* (not engine) turned lines. It is called
Uphanteria chemungensis, and is supposed to be a marine plant
(p. 183. Vanuxem). A fossil much like yours is figured by
De France in the *Dictionaire des Sciences Naturelles* under the
name of *Receptaculites neptuni,* from Chimay, in the Pays Bas.
This is certainly of the same genus. De Blainville also describes it in his *Actinologie* at the end of the corals, but offers
no opinion respecting its affinities. I should compare it with
Eschadites Konigi of Murchison's upper silurian, but *that* was
originally spherical and hollow."

bed of the river consists of limestone, sometimes lying in nearly horizontal layers more or less fissured ; in other places broken up into large loose slabs, tilted up and riding on each other. Boulders of granite occur in various parts of the river, some of them of considerable magnitude, and rising high out of the water. In the lower part of the river the banks are sandy, a considerable deposit of dry light soil overlies the limestone, and vegetation is early and vigorous.

When we left Lake Superior, in the middle of May, the deciduous trees gave little promise of life ; and, in ascending the Kamenistikwoya, we were glad to let the eye dwell upon the groves of aspens which skirt the streams in that undulating and rocky district, and which, when well massed, gave a pleasing variety to the wintry aspect of the landscape, — the silvery hue of their leafless branches and young stems contrasting well with the sombre green of the spruce fir, which forms the bulk of the forest. On the 27th of May, while ascending Church Reach of Rainy River, we had been cheered by the lovely yellowish hue of the aspens just unfolding their young leaves ; but the ice, lingering on Lake Winipeg, when we crossed it, had kept down the temperature, spring had not yet assumed its sway, and the trees were leafless. Now, the season seemed to be striding onwards rapidly, and the tender foliage was trembling on all sides in the bright sunshine. It was in a patch of burnt woods in this vicinity that, in the year 1820, I discovered the beautiful *Eutoca Franklinii*, now so common an ornament of our gardens.

Constantly, since the 1st of June, the song of the *Fringilla leucophrys* has been heard day and night, and so loudly, in the stillness of the latter season, as to deprive us at first of rest. It whistles the first bar of " Oh dear, what can the matter be ! " in a clear tone, as if played on a piccolo fife ; and, though the distinctness of the notes rendered them at first very pleasing, yet, as they haunted us up to the Arctic circle, and were loudest at midnight, we came to wish occasionally that the cheerful little songster would time his serenade better. It is a curious illustration of the indifference of the native population to almost every animal that does not yield food or fur, or otherwise contribute to their comfort or discomfort, that none of the Iroquois or Chippeways of our company knew the bird by sight, and they all declared boldly that no one ever saw it. We were, however, enabled, after a little trouble, to identify the songster, his song, and breeding-place. The nest is framed of grass, and placed on the ground under shelter of some small inequality ; the eggs, five in number, are greyish- or purplish-white, thickly spotted with brown; and the male hides himself in a neighbouring bush while he serenades his mate.

At the outlet of Beaver Lake, and at several succeeding points on both sides of the canoe-route, the thin slaty limestone forms cliffs, thirty or forty feet high; but about the middle of the lake, there is a small island of greenstone. Beyond this we again touched upon the granite rocks which we had left at the north-east corner of Lake Winipeg, bearing from this place about east 82° south.

At the entrance of Ridge River we met Mr. M'Kenzie, Jun., in charge of a brigade of boats, carrying out the furs of the Isle à la Crosse district, and were glad to obtain from him tidings of Mr. Bell, who was advancing prosperously, though he had been stopped for three days by ice, on the lake which we had just crossed. The Missinipi, or Churchill River, Mr. M'Kenzie told us, did not open till the 6th of the present month, though in common years it seldom continues frozen beyond the 1st.

Soon after parting with this gentleman, we met the schoolmaster of Lac La Ronge district, who, with his wife and four children, were on their way to pass the summer with the Rev. Mr. Hunter, episcopal clergyman at the Pas. Both husband and wife are half-breeds, and both are lively, active, and intelligent. The family party were travelling in a small canoe, which the husband paddled on the water, and carried over the portages with their light luggage. For their subsistence, they depended on such fish and wild-fowl as they could kill on the route; and the lady was very grateful for a small supply of tea, sugar, and flour which we gave her. The young ones bore the assaults of the moschetoes with a stoical indifference, as an inevitable evil, that had belonged to every summer of their lives, and from which no part of the world, as far as they knew, was exempt. At the Ridge Portage, where we encamped for the night, the rock is gneiss, resembling mica-slate, owing to the quantity of mica that enters into its composition.

On the 17th, we came early to a long and strong rapid, bearing the same appellation with the preceding portage, and which is said to be the highest point to which sturgeon ascend in this river; and it is most probably the northern limit of the range of that fish, on the east side of the Rocky Mountains. It is situated in about 54½° degrees of north latitude. We noticed two species of this fish in the Saskatchewan River system. One of these is described in the *Fauna Boreali-Americana* under the name of *Accipenser rupertianus*, and has a tapering acute snout. It seldom exceeds ten or fifteen pounds in weight. The other is the *Namè-yu* of the Crees, and has not been hitherto described. It very commonly weighs ninety pounds, and attains the weight of one hundred and thirty. Its snout is short and blunt, being only one third of the length of the entire head ; its nasal barbels are short, its shields small and remote, and the ventral rows are absent. Its caudal is less oblique than that of the smaller kind, the upper lobe being proportionally shorter. This species ascends the Winipeg River as high as the outlet of Rainy Lake : and the smaller kind is occasionally, though rarely, taken also in that locality, but, in general, it seems to be unable to surmount the cascade at the outlet of the Lake of the Woods. The rocks here are granite, and a mountain-green chlorite slate, similar to that which occurs so abundantly on the north side of the Lake Superior basin ; the latter, under the action of the weather, forms a tenacious clayey soil. A hornblende-slate occupies the bed of the river, and rises, on each bank, into rounded knolls and low cliffs. The inequalities of the country here, as well as its vegetation, are very similar to that on the Kamenistikwoya, where the same formation exists.

The woods, being now in full but still tender foliage, were beautiful. The graceful birch, in particular, attracted attention by its white stem, light green spray, and pendent, golden catkins. Willows of a darker foliage lined the river bank; and the background was covered with dark green pines, intermixed with patches of lively aspen, and here and there a tapering larch, gay with its minute tufts of crimson flowers, and young pale green leaves. The balsam poplar, with a silvery foliage though an ungainly stem, and the dank elder, disputed the strand at intervals with the willows; among which the purple twigs of the dog-wood contributed effectively to add variety and harmony to the colours of spring.

The *Actæa alba* grows abundantly here; it is called by the Canadians *le racine d'ours*, and by the Crees, *musqua-mitsu-in* (bears' food). A decoction of its roots and of the tops of the spruce fir is used as a drink in stomachic complaints. The *Acorus calamus* is another of the indigenous plants that enter into the native pharmacopœia, and is used as a remedy in colic. About the size of a small pea of the root, dried before the fire or in the sun, is a dose for an adult, and the pain is said to be removed soon after it is masticated and swallowed. When administered to children, the root is rasped, and the filings swallowed in a glass of water, or of weak tea with sugar. A drop of the juice of the recent root is dropped into inflamed eyes, and the remedy is said to be an effectual though a painful one. I have never seen it tried. The Cree name of this plant is *watchŭskĕ mitsu-in*, or " that which the musk-rat eats."

At breakfast-time we crossed the Carp Portage, where there is a shelving cascade over granite rocks. The grey sucking carp (*Catastomus hudsonius*) was busy spawning in the eddies, and our voyagers killed several with poles. Two miles above the portage there are some steeply rounded sandy knolls clothed with spruce trees, being the second or high bank of the river, which is elevated above all floods of the present epoch. In some places granite rocks show through sand, heaped round their base. The frequent occurrence of accumulations of sand in this granite and gneiss district, near the water-sheds of contiguous river systems, has been already noticed. In the course of the forenoon we passed the Birch lightening-place (*Demi-charge du bouleau*), where a slaty sienite or greenstone occurs, the beds being inclined to the east-north-east at an angle of 45°; and an hour afterwards we crossed the Birch Portage, five hundred and forty paces long. The rocks there are porphyritic granite, portions of which are in thin beds, and are therefore to be entitled gneiss.

The river has the character peculiar to the district,

that is, it is formed of branching lake-like expansions without perceptible current, connected by falls or rapids occasioning portages, or by narrow straits through which there flows a strong stream. At four in the afternoon we crossed the Island Portage, where the rock is a fine-grained laminated granite or gneiss, containing nodules or crystals of quartz, which do not decay so fast as the rest of the stone, and consequently project from its surface: the layers are contorted. In 1825, which was a season of flood, this islet was under water, and our canoes ascended among the bushes.

Two hours later we passed the Pine Portage (*Portage des Epinettes*), and entered Half-Moon Lake (*Lac Mi-rond**). At this portage the rocks are granite, greenstone, and black basalt, or hornblende-rock, containing a few scales of mica, and a very few garnets. The length of the portage is two hundred yards. At our encampment on a small island in Half-Moon Lake the gneiss lay in vertical layers, having a north and south strike. A few garnets were scattered through the stone. This piece of water, and Pelican and Woody Lakes, which adjoin it, are full of fish, and they are consequently haunted by large bodies of pelicans, and several pairs of white-headed eagles (*Haliæetus*

---

* Called by mistake *Lac Heron* in Franklin's overland journeys.

albicilla). This fishing eagle abounds in the watery districts of Rupert's Land; and a nest may be looked for within every twenty or thirty miles. Each pair of birds seems to appropriate a certain range of country on which they suffer no intruders of their own species to encroach; but the nest of the osprey is often placed at no great distance from that of the eagle, which has no disinclination to avail itself of the greater activity of the smaller bird, though of itself it is by no means a bad fisher. The eagle may be known from afar, as it sits in a peculiarly erect position, motionless, on the dead top of a lofty fir, overhanging some rapid abounding in fish. Not unfrequently a raven looks quietly on from a neighbouring tree, hoping that some crumb may escape from the claws of the tyrant of the waters. Some of our voyagers had the curiosity to visit an eagle's nest, which was built, on the cleft summit of a balsam poplar, of sticks, many of them as thick as a man's wrist. It contained two young birds, well fledged, with a good store of fish, in a very odoriferous condition. While the men were climbing the tree the female parent hovered close round, and threatened an attack on the invaders; but the male, who is of much smaller size, kept aloof, making circles high in the air. The heads and tails of both were white.

The pelican, as it assembles in flocks, and is

very voracious, destroys still larger quantities of
fish than the eagle. It is the *Pelicanus trachy-
rhynchus* of systematic ornithologists, and ranges
as far north as Great Slave Lake, in latitude 60°—
61° N. These birds generally choose a rapid for
the scene of their exploits, and, commencing at
the upper end, suffer themselves to float down with
the current, fishing as they go with great success,
particularly in the eddies. When satiated, and
with full pouches, they stand on a rock or boulder
which rises out of the water, and air themselves,
keeping their half-bent wings raised from their
sides, after the manner of vultures and other gross
feeders. Their pouches are frequently so crammed
with fish that they cannot rise into the air until
they have relieved themselves from the load, and
on the unexpected approach of a canoe, they stoop
down, and, drawing the bill between their legs,
turn out the fish. They seem to be unable to accomplish this feat when swimming, so that then
they are easily overtaken, and may be caught alive,
or killed with the blow of a paddle. If they are near
the beach when danger threatens, they will land to
get rid of the fish more quickly. They fly heavily,
and generally low, in small flocks of from eight to
twenty individuals, marshalled, not in the cunei-
form order of wild geese, but in a line abreast,
or slightly *en echellon ;* and their snow-white plum-

age with black-tipped wings, combined with their great size, gives them an imposing appearance. Exceeding the fishing eagle and the swan in bulk, they are the largest birds in the country. Their eggs are deposited on rocky islets among strong rapids, where they cannot be easily approached by man or beasts of prey. The species is named from a ridge or crest which rises from the middle line of the upper mandible of the male; sometimes from its whole length, when it is generally uneven; and sometimes from a short part only, when it is semicircular and smooth-edged.

The black-bellied tern (*Hydrochelidon nigra*) is also abundant on these waters, and ranges northwards to the upper part of the Mackenzie. And the Cayenne tern (*Sterna cayana*) is common in this quarter and onwards to beyond the arctic circle; but notwithstanding Mr. Rae's expertness as a fowler, and eagerness to procure me a specimen, the extreme wariness of the bird frustrated all his endeavours until this day, when he brought one down, and gave me an opportunity of examining it, which I was glad to do, since from want of a northern specimen the bird was not noticed in the *Fauna Boreali-Americana*. Mr. Audubon mentions the great difficulty of shooting this bird, and he succeeded in doing so only by employing several boats to approach its haunts in different directions.

Albert, our Eskimo interpreter, told me that it
does not visit Hudson's Bay.

I was also indebted to Mr. Rae's gun for specimens of the brown crane (*Grus canadensis*).
Mr. Audubon, who is so competent an authority
on all questions relating to American birds, and
whose recent death all lovers of natural history
deplore, was of opinion, in common with many
other ornithologists, that the brown crane is merely
the young of the large white crane (*Grus americana*); but, though I concede that the young of the
latter are *grey*, I think that the *brown* species is
distinct; first, because it is generally of larger
dimensions than the white bird, and secondly, because it breeds on the lower parts of the Mackenzie
and near the Arctic coasts, where the *Grus americana* is unknown.   As far as I could ascertain,
the latter bird does not go much further north
than Great Slave Lake.   At *Fort aux Liards*, on
the River of the Mountains, large flocks of cranes
pass continually to the westward, from the 17th
to the 20th of September; the grey and white birds
being in different bands, and the former of smaller
size, like young birds.   Very rarely during the
summer a flock of white cranes passes over Fort
Simpson in latitude 62° N.   The brown cranes, on
the other hand, which frequent the banks of the

Mackenzie from Fort Norman, in latitude 65° N. down to the sea-coast, are generally in pairs. They are in the habit of dancing round each other very gracefully on the sand-banks of the river.*

June 18th. — About three hours after embarking we came to the Pelican lightening-place (*Demi-charge de chetauque*), and by breakfast-time we had crossed the three portages of Woody Lake. A micaceous gneiss or mica-slate rock prevails at these portages. A family of Cree Indians, who were encamped on one of the many islands which adorn the scenery of Woody Lake, exchanged fish for tobacco, and enabled us to vary the diet of our voyagers, an indulgence which pleases them greatly; for, though they generally prefer pork or pemican to fish, they relish the latter occasionally. At five we crossed the Frog Portage, or *Portage de Traitè* of the Canadians, and encamped on the banks of the Missinipi or Churchill River, in the immediate vicinity of a small outpost of the Hudson's Bay Company.†

* Much of this information I received from Murdoch Mac Pherson, Esq., who, during twenty years' residence on the Mackenzie, became thoroughly acquainted with all its feathered and ferine inhabitants.

† The Cree term *Missinipi* signifies "much water," and is analogous to that of *Mississippi*, which means "great river;" *nipi* being water, and *sipi* river. The Canadians call it English

No change of formation takes place in passing from the Saskatchewan River system to that of the Missinipi. The Frog Portage is low, and Churchill River, in seasons of flood, sometimes overflows it and discharges some of its superfluous waters into the Woody Lake.* The general level of the country for some distance, or down to the lower end of Half-Moon Lake, varies little; but in this and in Pelican Lake there are a few conical eminences, which rise several hundred feet above the water. We did not approach them sufficiently near for examination.

Frog Portage is the most northerly point of the Saskatchewan basin, and lies in 55° 26′ N. latitude, 103° 20′ W. longitude. Below it there is a remarkable parallelism in the courses of Churchill and Saskatchewan rivers, both streams inclining to the north-east in their passage through the " intermediate primitive range," the district from whence they receive lateral supplies being at the same time very greatly narrowed. Several other considerable

River, because on it the early fur traders from Canada encountered the Hudson's Bay Company's people ascending from their principal depôt at Fort Churchill.

* About forty years ago, in a season remembered especially for the land-floods, a gentleman was drowned on the Frog Portage by his canoe oversetting against a tree as he was passing from Churchill River.

streams run near them, and parallel to them, but do not originate so far to the westward. In their widely spreading upper branches, and their restricted trunks, they resemble trees. As they are not separated high on the prairie slopes by an elevated water-shed, they may be considered, in reasoning upon the direction of the force which excavated their basins, as one great system, having an eastern direction and outlet, interposed between the Missouri and Mackenzie, which discharge themselves respectively into southern and northern seas.

The Churchill River is the boundary between the Chepewyans and Crees, but a few of the latter frequent its borders, resorting to Lac la Ronge and Isle à la Crosse posts, along with the Chepewyans, for their supplies.

On June the 19th, a fog detained us at our encampment to a later hour than usual; when being unwilling to lose all the morning, we went some distance in the thick weather under the guidance of the post-master, who was acquainted with every rock in the neighbourhood. As the sun rose higher the atmosphere cleared, and we ascended the Great Rapid by its southern channel, making a portage part of the way, and poling up the remainder. A recent grave with its wooden cross marked the burial-place of one of the Hudson's Bay Company's servants, who was drowned here last year. His body was thrown out a little below the rapid.

We next crossed the Rapid lightening-place, and afterwards mounted four several rapids, connected with the Barrel Portage. In the afternoon the Island Portage was made, where the river, being pent in for a short space between high, even, rocky banks, is there only five or six hundred yards wide, and has a strong current, requiring much exertion from the canoe-men in paddling round the headlands. Elsewhere, except at the rapids during this day's voyage, Churchill River has more the character of a lake. In the evening we crossed the portage of the Rapid River, one hundred and sixty paces long, which has its name from a tributary stream on which the Hudson's Bay Company have a post, that is visible from the canoe-route. Afterwards we passed the lightening-place of the Rapid River, and encamped five or six miles further on, at half-past eight o'clock.

Our Iroquois, being tired with the day's journey and longing for a fair wind to ease their arms, frequently in the course of the afternoon, scattered a little water from the blades of their paddles as an offering to *La Vieille*, who presides over the winds. The Canadian voyagers, ever ready to adopt the Indian superstitions, often resort to the same practice, though it is probable that they give only partial credence to it.   Formerly the English shipmen, on their way to the White Sea, landed regularly in Lapland to purchase a wind from the witches residing near North Cape; and the rudeness and fears of Frobisher's companions in plucking off the boots or trowsers of a poor old Eskimo woman on the Labrador coast, to see if her feet were cloven, will be remembered by readers of arctic voyages.

Throughout the day's voyage, the primitive formation continued.   In several places we observed micaceous slate, traversed by large veins of granite, and alternating with beds of the same, also gneiss in thick beds, with its layers much contorted. Below the Great Rapid there are many bluff granite rocks, and some precipices thirty or forty feet high, the higher knolls rising probably from two to three hundred feet above the water.   At the Great Rapid a greenstone-slate stained with iron occurs.   At the Barrel Portage, a mile or two further on, where the river makes a sharp bend, beds of chlorite-slate occupy its channel for two miles, having a north-east and south-west strike, and a southerly dip of 60° or 70°.   Beds of greenstone-slate are interleaved with it.   Above the Island Portage a sienite occurs which contains an imbedded mineral; and at the Rapid River Portage, mica-slate, passing into gneiss, prevails, the beds having a south-west and north-east strike. The granite veins here have a general direction nearly coincident with that of the beds, but they are waved up and down. In the vicinity of the veins the layers of slate are much contorted, following the curvatures of the veins closely. At the lightening-place of the Rapid River, there is a fine precipice of granite fifty feet high, which is traversed obliquely from top to bottom by two magnificent veins of flesh-coloured porphyry-granite. Five miles further on there are precipices of granite one hundred and fifty feet high.

The country in this neighbourhood is hilly, and a few miles back from the river the summits of the eminences appear to the eye to rise four hundred or perhaps five hundred feet above the river. The resemblance of the whole district to Winipeg River is perfect, and the general aspect of the country is much like that of the north shores of Lake Superior, though the water basin is not so deeply excavated.

An hour and a half after starting on the morning of the 20th, we crossed the Mountain Portage, one hundred paces long, where the rock is hornblende-slate. At Little Rock Portage, a short way further on, the thin slaty beds have a north-east and south-west strike. Above this, a dilatation of the river, named Otter Lake, leads to the Otter

Portage of three hundred paces made over mica-slate. The beach there is strewed with fragments of a crystalline augitic greenstone, showing that that rock is not far distant.

From a party of Chepewyans who were encamped on the Otter Lake, we procured a quantity of a small white root, about the thickness of a goose quill, which had an agreeable nutty flavour. I ascertained that it was the root of the *Sium lineare*. The poisonous roots of *Cicuta virosa, maculata*, and *bulbifera*, are often mistaken for the edible one, and have proved fatal to several labourers in the Company's service. The natives distinguish the proper kind by the last year's stem, which has the rays of its umbel ribbed or angled, while the *Cicutæ* have round and smooth flower-stalks. When the plant has put out its leaves, by which it is most easily identified, the roots lose their crispness and become woody. The edible root is named *ūskotask* by the Crees, and *queue de rat* by the Canadians. The poisonous kinds are called *manito-skatask*, and by the voyagers *carrotte de Moreau*, after a man who died from eating them.

The *Heuchera Richardsonii*, which abounds on the rocks of this river, is one of the native medicines, its astringent root being chewed and applied as a vulnerary to wounds and sores. Its Cree name is *pichē quaow-utchēpi*. The leaves of the *Ledum palustre* are also chewed and applied to burns, which are said to heal rapidly under its influence. The cake of chewed leaves is left adhering to the sore until it falls off.

In the course of the forenoon we ascended four rapids, occasioning short portages, then the Great Devil's Portage, of fourteen hundred paces; and in the afternoon several other rapids were passed, among which were the Steep Bank, Little Rock, and Trout portages. At the Steep Bank Portage (*Portage des Ecores*), which is one hundred and sixty paces long, gneiss and mica-slate occur interleaved irregularly with each other, and intersected in every direction by reticulating quartz veins; the prevailing rock in the neighbourhood being gneiss, and the hills low and barren.

June 21st. — Soon after starting we crossed the Thicket Portage (*Portage des Haliers*) of three hundred and sixty paces, and entered Black-bear Islands Lake, a very irregular piece of water, intersected by long promontories and clusters of islands. After four hours' paddling therein we came to a rapid, considered by the guide as the middle of the lake; in three hours more we came to another strong rapid, and after another three hours to the Broken-Canoe Portage, which is at the upper end of this dilatation of the river. Granite is the prevailing rock in the lake, and one of the small islands consists of large balls of that stone, piled on each other like cannon shot in an arsenal. They might be taken for boulders were they not heaped up in a conical form and all of one kind of stone; and they have obviously received their present form by the softer parts of the rock having crumbled and fallen away. At Thicket Portage and the lower end of the lake, the granite is associated with greenstone slate; and at Broken-Canoe Portage, above the lake, a laminated stone exists, whose vertical layers are about an inch thick, and have a north and south strike, being parallel to the direction of the ridges of the rock. This stone is composed of flesh-coloured quartz, with thin layers of duck-green chlorite, and no felspar. It ought perhaps to be considered as a variety of gneiss.

Later in the afternoon we came to the Birch and Pin Portages, on the last of which we encamped. The granite rocks here are covered by a high bank of sand and gravel, filled with boulders.

June 22*d.* — Embarking early, we passed through Sand-fly Lake, and afterwards Serpent Lake, in which we met the Athabasca brigade of boats, under charge of Chief Trader Armitinger. This gentleman informed us that he met Mr. Bell with our boats on the 19th, on which day they would arrive at Isle à la Crosse. The aspect of the country changes suddenly on entering these lakes. The rising grounds have a more even outline, and one long low range rises over another, as the country recedes from the borders of the water, where it is generally low and swampy. The trees near the water are almost exclusively birch and balsam-poplar, or aspen ; the spruce firs occupying the distant elevations, which are generally long round-backed hills, with a few short conical bluffs. Serpent Lake is named from the occurrence on its shores of a small snake.* I was not able to learn that this or any other snake had been detected further to the north. Having passed a high sand-bank on the north side of Serpent Lake, six miles further on, we entered the Snake River, within the mouth of which there is a bank of loam, sand, and rolled stones, thirty-five feet high. The bed of the stream is lined with these stones, and its width is about equal to that of Rainy River. The rocky points, as seen from the canoe, appeared to be of granite. All the boulders that I examined were of a dull brownish-red, striped or laminated granite, which, on a cursory inspection, might be mistaken for sandstone. Boulders of the same kind occur at the Snake Rapid, where they are intermixed with a few pieces of hornblende rock.

* *Coluber* or *Tropidinotus sirtalis.*

June 23*d.*—The moschetoes were exceedingly numerous and troublesome during the night and this morning. Our route lay through Sandy Lake and Grassy River, where the country retains the same general aspect that it has on Sand-fly and Serpent Lakes, and where the prevailing rock is a brownish-red, fine-grained sienite, resembling a sandstone. The same rock abounds in Knee Lake, where, however, we saw, for the first time since leaving the south end of Lake Winipeg, fragments of white quartzose sandstone; but did not find the stone *in situ.* The sienite, when traced, is found to pass into hornblendic granite, by the addition of scales of mica to some parts of the same beds. The high banks of Knee Rapid consist of sandy loam crammed with boulders.

The *Tabanus*, named by the voyagers " Bull-dog," has been common for two days. The current notion is, that this fly cuts a piece of flesh from his victim, and at first sight there seems to be truth in the opinion. The fly alights on the hands or face so gently that if not seen he is scarcely felt until he makes his wound, which produces a stinging sensation as if the skin had been touched by a live coal. The hand is quickly raised towards the spot, and the insect flies off. A drop of blood, oozing from the puncture, gives it the appearance of a gaping wound, and the fly is supposed to have carried off a morsel of flesh. In fact, the *Tabanus* inserts a five-bladed lancet, makes a perforation like a leech-bite, and, introducing his flexible proboscis, proceeds to suck the blood. He is, however, seldom suffered to remain at his repast; unless, as in our case, he be allowed to do so, that his mode of proceeding may be inspected. These *Tabani* are troublesome only towards noon and in a bright sun, when the heat beats down the moschetoes.*

In the afternoon we passed through Primeau's Lake, having previously ascended three strong and bad rapids. At the middle turn of the lake a moderately high, long, and nearly level-topped hill closes the transverse vista. The channel between the eastern and western portions of the lake winds among extensive sandy flats, covered with bents, and in some places there was a rich crop of grass not in flower, but seemingly a *Poa*. In the evening we encamped at the " Portage of the Exhausted," on the river between Isle à la Crosse and Primeau Lakes. The rock here, and on the two lakes below it, is the brownish-red slaty sienite already mentioned : it has much resemblance to a

* Of the five lancets with which the *Tabanus* wounds his prey two are broader than the others. They are enclosed in a black hairy sheath, whose extremity folds back. The palpi are conico-cylindrical and tubular.

rock on Lakes Huron and Superior, which seemed there to be associated with a conglomerate. The brownish colour belongs to the felspar; a vitreous quartz also enters into its composition, and a little hornblende. It is rather easily frangible, and has a flat, somewhat slaty, fracture.

Two hours after embarking on the 24th we passed the Angle Rapid (*Rapide de l'Equerre*), and subsequently the Noisy (*Rapide Sonante*), and Saginaw Rapids, and entered the small Saginaw Lake, which we crossed in half an hour. At various points we had cursory glances, in passing, of granite forming low rocks. The Crooked Rapid, a mile and a half long, conducted us to Isle à la Crosse Lake. In traversing twenty-three geographical miles of this lake, we disturbed many bands of pelicans, which were swimming on the water, or seated on rocky shoals, in flocks numbering forty or fifty birds. On the shores there are fragments of a white quartzose sandstone, but I noticed no limestone. The country consists of gravelly plains, having a coarse sandy soil and numerous imbedded boulder stones. Shoals formed by accumulations of boulders are common in the lake, and in various places close pavements of these stones are surmounted by sandy cliffs twenty or thirty feet high. The bulk of the boulders belongs to the brownish glassy sienite mentioned in a preceding page.

The funnel-shaped arm named Deep River (*La Rivière Creuse*) meets the northern point of the lake at an acute angle, enclosing between it and Clear Lake a triangular peninsula. Beaver River, the principal feeder of the lake, flows from Green Lake, which lies directly to the southward, near the valley of the Saskatchewan in the 54th parallel of latitude. The winter path from Isle à la Crosse to Carlton House ascends this river to its great bend, whence it leads to the Saskatchewan plains, through an undulating country but without any marked acclivity. I consider it probable, therefore, that Isle à la Crosse Lake and Carlton House do not differ more than two hundred feet from each other in their height above the sea. The altitude of the latter I have judged to be about eleven hundred feet; and Captain Lefroy, from his experiments with the boiling-water thermometer, assigns an elevation of thirteen hundred feet to the former.

Churchill River, disregarding its flexures, has a course to the sea from Isle à la Crosse Fort of five hundred and twenty-five geographical miles, and the length of the Saskatchewan below Carlton House is six hundred and thirty miles. The general descent of the eastern slope of the con-

tinent to Hudson's Bay from these two localities
may be reckoned at a little more than two feet a
mile.   Further to the westward, in the vicinity of
Fort George, near the 110th meridian, the upper
branches of the Beaver River rise from the very
banks of the Saskatchewan.

On Beaver River the strata are limestone, and
a line drawn from the north side of Lake Winipeg,
to the south side of Isle à la Crosse Lake, runs about
north 58° west, and touches upon the northern edge
of the limestone in Beaver Lake.   That line may,
therefore, be considered as representing the general
direction of the junction of the limestone with the
primitive rocks in this district of the country.
Judging from relative geographical position and
mineralogical resemblances, the north part of Isle à
la Crosse Lake belongs to a similar sandstone deposit
with that which skirts the primitive rocks on Lake
Superior, — a peculiar looking sienite being connected with the sandstone in both localities.   From
its order of occurrence the limestone of Beaver River
is probably silurian.   My observations were too
limited and cursory to carry conviction, even to my
own mind, on these points; the circumstances attending the several journeys I have made through
these countries having prevented me from obtaining
better evidence.   In a voyage with ulterior objects
through so wide an extent of territory, and with so

short a travelling season, every hour is of importance, and whoever has charge of a party must
show that he thinks so, otherwise his men cannot
be induced to keep up their exertions for sixteen
hours a day, which is the usual period of labour in
summer travelling.   Of this time an hour's halt is
allowed for breakfast, and half an hour for dinner.
We did not reach Isle à la Crosse Fort till half-
past nine in the evening, and then learnt that Mr.
Bell with the boats was four days in advance of us.

June 25th.— A strong gale blowing this morning
detained us at the post, and the day being Sunday
our voyagers went to mass at the Roman Catholic
chapel, distant about a mile from the fort.   This
mission was established in 1846 under charge of
Monsieur La Flêche, who has been very successful
in gaining the confidence of the Indians, and
gathering a considerable number into a village
round the church.   In the course of the day I
received a visit from Monsieur La Flêche and his col-
league Monsieur Taschè.   They are both intelligent
well-informed men, and devoted to the task of in-
structing the Indians; but the revolution in France
having cut off the funds the mission obtained from
that country, its progress was likely to be impeded.
They spoke thankfully of the assistance and countenance they received from the gentlemen of the
Hudson's Bay Company.   The character they gave

the Chepewyans for honesty, docility, aptness to receive instruction, and attention to the precepts of
their teachers, was one of almost unqualified praise,
and formed, as they stated, a strong contrast to that
of the volatile Crees. They have already taught
many of their pupils here to read and write a
stenographic syllabic character, first used by the
late Reverend Mr. Evans, a Wesleyan missionary,
formerly resident at Norway House, but which
Monsieur La Flêche has adapted to the Chepewyan
language. On asking this gentleman his opinion
of the affinity between the Cree and Chepewyan
tongues, both of which he spoke fluently, he told me
that the grammatical structure of the Chepewyan
was different, the words short, and the sounds dis-
similar, bearing little resemblance to the soft, flowing compounds of the Cree language.

As there is generally some difficulty in making
an early start from a fort, we moved in the evening
to the point of the bay, that we might be ready to
take advantage of the first favourable moment for
proceeding on our voyage.

June 26th. — We embarked before 3 A. M., but a
strong head-wind blowing, we could proceed only
by creeping along-shore under shelter of the projecting points. For some days past the water has
been covered with the pollen of the spruce fir, and
to-day we observed that it was thickly spread with

the downy seeds of a willow. The banks of Deep River, which forms the discharge of Buffalo and Clear Lakes, consist of gravel and sand containing large boulders, principally of trap and primitive rocks. The eminences rise from fifteen to forty feet above the river, and the land-streams have cut ravines into the loose soil, the whole being well covered with the ordinary trees of the country. This low land extends to Primeau Lake on the one side, and Buffalo Lake on the other. The beach, especially towards the openings of Cross and Buffalo Lakes, is strewed with fragments of quartzose sandstone, mixed with some pieces of light-red freestone, and many boulders of earthy greenstone, chlorite-slate, porphyritic greenstone slate, and gneiss. Neither mica-slate nor limestone were observed among them, and no rocks *in situ*. Many of the bays have sandy beaches. The Deep River has little current, except where it issues from the lakes.

In the morning a Canada lynx was observed swimming across a strait, where the distance from shore to shore exceeded a mile. We gave chase, and killed it easily. This animal is often seen in the water, and apparently it travels more in the summer than any other beast of prey in this country. We put ashore to sup at seven in the evening, at a point in Buffalo Lake, where we found evidences of the

boat party having slept there a night or two previously.   Being desirous of overtaking them without delay, we immediately resumed our voyage,
but were caught in the middle of the lake by a
violent thunder-storm, accompanied by strong
gusts of wind.   The voyagers were alarmed, and
pulled vigorously for the eastern shore, on which
we landed soon after eleven.   The shores of
Buffalo Lake are generally low; but, on the west
side, there is an eminence named Grizzle Bear
Hill, which is conspicuous at a considerable distance.   It probably extends in a north-west direction towards the plateau of Methy Portage and
Clear-water River.   The valley to the east is
occupied by Methy, Buffalo, and Clear Lakes, the
last of which is said to have extensive arms.

Embarking at daylight on the 27th, we crossed
the remainder of the lake, being about fourteen
miles, and entered the Methy River, which we
found to our satisfaction higher than usual; as in
so shallow a stream the navigation is very tedious
in dry seasons.   The watermarks on the trees
skirting the river showed that the water had
fallen at least five feet, since the spring floods.
The moschetoes are more numerous in seasons of
high water, and this year was no exception to the
general rule.

At the Rapid of the Tomb (*La Cimetiere*) several

pitch or red pines (*Pinus resinosa*) grow intermixed with black spruces, one of them being a good-
sized tree. This is the most northerly situation in
which I saw this pine, and the voyagers believe that
it does not grow higher than the River Winipeg.

An Indian, who has built a house at the mouth
of the river, keeps fifteen or twenty horses, which
he lets to the Company's men on Methy Portage,
the charge being " a skin," or four shillings, for
carrying over a piece of goods or furs weighing
ninety pounds. From him we received the very
unpleasant intelligence, that not only had his
horses died of murrain last autumn, but that all
the Company's stock employed on the portage had
likewise perished. This calamity foreboded a detention of seven or eight days longer on the
portage than we expected, and a consequent re-
duction of the limited time we had calculated upon
for our sea-voyage. I had used every exertion to
reach the sea-coast some days before the appointed
time, expecting to be able to examine Wollaston's
Land this season; — this hope was now almost extinguished. Another stock of horses had been ordered from the Saskatchewan, but they were not
likely to arrive till the summer was well advanced.

Methy River flows through a low, swampy country, of which a large portion is a peat moss. Some
sandy banks occur here and there, and boulders

are scattered over the surface, and line the bed of the stream. We encamped on the driest spot we could find, and had to sustain the unintermitting attacks of myriads of moschetoes all night.

The Methy River, Lake, and Portage, are named from the Cree designation of the Burbot (*Lota maculosa*) (*La Loche* of the Canadians), which abounds in these waters, and often supplies a poor and watery food to voyagers whose provisions are exhausted. Though the fish is less prized than any other in the country, its roe is one of the best, and, with a small addition of flour, makes a palatable and very nourishing bread.

Four hours' paddling brought us, early on the 28th, to the head of the river, and two hours more enabled us to cross to the eastern side of Methy Lake, where we were compelled to put ashore by a strong headwind. A female mink (*Vison lutreola*) was killed as it was crossing a bay of the lake. It had eight swollen teats, and its udder contained milk ; so that probably its death ensured that of a young progeny also. The feet of this little amphibious animal are webbed for half the length of its toes. It is the *Shakwèshew* or *Atjakashew* of the Crees, the "Mink" of the fur-traders, and the *Foutereau* of the Canadians.

In the evening, the wind having decreased, we paddled under shelter of the western shore to the upper end of the lake, and entered the small creek which leads to the portage.

Mr. Bell was encamped at the landing-place, having arrived on the previous day, which he had spent in preparing and distributing the loads, and the party had advanced one stage of different lengths, according to the carrying powers of the individuals, which were very unequal. On visiting the men, I found two of the sappers and miners lame from the fatigue of crossing the numerous carrying-places on Churchill River, and unfit for any labour on this long portage. Several others appeared feeble ; and, judging from the first day's work of the party, I could not estimate the time that would be occupied, should they receive no help in transporting the boats and stores, at less than a fortnight, which would leave us with little prospect of completing our sea-voyage this season. In the equal distribution of the baggage each man had five pieces of ninety pounds' weight each, exclusive of his own bedding and clothing, and of the boats, with their masts, sails, oars, anchors, &c., which could not be transported in fewer than two journeys of the whole party. The Canadian voyagers carry two pieces of the standard weight of ninety pounds at each trip on long portages such as this, and, in shorter ones, often a greater load. Several of our Europeans carried only one piece at a time, and had, consequently, to make five trips with their share of the baggage, besides two with the boats; hence they were unable to make good the ordinary day's journey of two miles, being, at seven trips with the return, twenty-six miles of walking, fourteen of them with a load. The practised voyager, on the contrary, by carrying greater loads, can reduce the walking by one third, and some of them by fully one half.*

* In 1825 Sir John Franklin ascertained the position of the first resting-place, after leaving Methy Lake, to be in latitude 56° 36′ 30″ N., longitude 109° 52′ 54″ W. By carefully pacing the distance from thence to Methy Lake, I found it to be 1790 yards, on a south 43° 25′ east bearing, giving 22″ difference of latitude, and 58″ of longitude. Hence the east end of the portage lies in latitude 56° 36′ 08″ N., longitude 109° 51′ 56″ W.

The usual encampment by the tomb on the south side of the Little Lake is in latitude 56° 40′ 17″ N., longitude 109° 57′ 54″ W., and the north end of the path on the banks of the Clearwater River is in 56° 42′ 51″ N., 109° 59′ 08″ W. The direct distance from one end of the portage to the other is therefore only 7½ geographical miles on a north 27° west course; while the paces, reduced to yards at the rate of 23 feet to every 10 paces (which I found after several trials to be the average), are 18,855, or 10·7 statute miles.

I subjoin the voyagers' names for the several resting-places on the portage, premising, however, that the halting-places vary both in number and position with the loads and strength of the carriers, and that the names are often transposed.

Methy Lake (*Lac la Loche*).

Thence to *Petit Vieux*	-	-	-	2557 paces.
,, *Fontaine du Sable*	-	-	-	3171 ,,
,, *La Vieille*	-	-	-	4591 ,,

By their agreements, our canoe-men were at liberty to return as soon as we overtook the boats; and, in that case, the additional pieces we had brought would of course be added to the baggage of the boat party; but I engaged them to assist us during the time that we were occupied on the portage, for an increase of wages of four shillings, York currency, per diem each.

June 29th. —— Our canoe-men were early astir this morning, and, before breakfast-time, had carried all the cargo of the canoes to the banks of a small lake, being two thirds of the whole portage, or 16,724 paces: the entire distance from Methy Lake to Clear-water River is 24,593 paces.

By observations with the aneroid and Delcros' barometers, I ascertained that the Little Lake was elevated twenty-two feet above Methy Lake; that the highest part of the pathway between the Little Lake and the Clear-water River rises above the latter six hundred and fifty-six feet, but, above Methy Lake, only sixty-six feet. The Cockscomb,

Thence to	*Bon Homme ou De Cyprès*	-	3167	paces.
„	*Petit Lac*	-	3238	„
„	*De Cyprès ou La Vieux*	-	4302	„
„	*La Crête*	-	1283	„
„	*Descente de la Crête*	-	1984	„
„	*La Prairie*	-	300	„
			24,593	„

or the crest of the precipitous brow which over-
looks the magnificent valley of the Clear-water, is
twenty-two feet lower than the summit of the path,
or six hundred and thirty-four feet above the last-
named river.   The portage-road is, in fact, nearly
level; the inequalities being of small account as far
as to the sudden descent of the Cockscomb.   In
the sandy soil there are many fragments of sandstone, a few of limestone, and scattered boulders of
granites, sienites, and greenstones.   The deposit of
sand is about six hundred feet deep, and most
probably encloses solid beds of sandstone.   It is
based on a (Devonian ?) limestone, which lines the
whole bed of the Clear-water River, till its junction
with the Elk River, as I shall hereafter mention.*

Captain Lefroy assigns fifteen hundred feet as
the elevation of the surface of Methy Lake above
the sea, and, from various estimates of the rate of
descent of Mackenzie River and its feeders, I am
inclined, independent of his calculations, to consider the Clear-water River at Methy Portage to
be nine hundred feet above the sea, which accords
well with his conclusions; since the difference of
level between Methy Lake and Clear-water River
being five hundred and ninety feet by my baro-

* As the Coxscomb is under the level of the brow of the
valley, the depth of sand may be more than 600 feet at its
highest points.

VOL. I.   I

metrical observations, the latter would be nine hundred and fifty feet above the sea by his data.[*]

[*] The exact height assigned by Captain Lefroy to Methy Lake is 1540 feet, which I have reduced in the text to the even number of 1500, as agreeing better with my own estimates. If this be nearly correct, Captain Lefroy gives too small an altitude to Isle à la Crosse Lake, since the route from thence to the portage is chiefly lake-way; and the Methy River cannot have a descent of 240 feet, which his altitudes would assign to it.

In the year 1848 I made several observations with the aneroid on Methy Portage to ascertain its levels, but they were neither so carefully made nor so extensive as they would have been, had I been less anxiously and constantly employed about the transport of the goods and boat. The error in this case is not, however, likely to be many feet, as the portage is evidently very nearly level as far as the Cockscomb. The height of the latter was ascertained on July 27. 1849, by Delcros' barometer, the observations being as follow : —

	Hour. A. M.		Delcros' barom. Millimr.	+0·34 cor. for general error.	Red. to Eng. inches.	Red. to temp. 32°	Att. Therm. Centr.	Fab.	Det. Th.
	h.	m.							
Six feet above Clear-water R.	4	0	72·719	72·753	28·644	28·606	6·4	43·5	40·8
Two feet above Cockscomb	4	46	71·079	71·113	27·998	27·944	10·2	50·4	50·9
Six feet above Clear-water R.	5	20	72·740	72·774	28·652	28·591	11·4	52·5	51·0

These furnish two sets for calculation,—

the first giving a height of	- - 640 feet
and the second of	- - - 632 ,,
The aneroid barometer in 1848 gave	- 631 ,,
Mean	- 634 ,,

Sir Alexander Mackenzie estimated this declivity at 1000 feet, Lieutenant Hood at 900 feet, both judging merely from the eye and time employed in its descent.

On the 3d of July, the whole of the baggage and the boats were brought to the banks of the Little Lake; and on the 6th, every thing having been carried over to Clear-water River on the preceding evening, we descended from the Cockscomb, where we had remained encamped for two days, that we might avoid the moschetoes which infested the low grounds. While the boats were loading, we took leave of our canoe-men, who returned to Canada, and at half-past eight A.M. we pushed off.

The portage occupied nine days from the time of Mr. Bell's arrival; but, with the assistance of horses, we could have passed it easily in three, and saved nearly a week of summer weather, most important for our future operations, besides husbanding the strength of the men. The transport of the four boats, being made on the men's shoulders, employed two days and a half of our time.

CHAP. IV.

CLEAR-WATER RIVER. — VALLEY OF THE WASHAKUMMOW. — PORTAGES. — LIMESTONE CLIFFS. — SHALE. — ELK OR ATHABASCA RIVER. — WAPITI. — DEVONIAN STRATA. — GEOLOGICAL STRUCTURE OF THE BANKS OF THE RIVER. — ATHABASCA LAKE, OR LAKE OF THE HILLS. — MEET MR. M'PHERSON WITH THE MACKENZIE RIVER BRIGADE. — SEND HOME LETTERS. — L'ESPERANCE'S BRIGADE. — FORT CHEPEWYAN. — HEIGHT OF LAKE ATHABASCA ABOVE THE SEA. — ROCKS. — PLUMBAGO. — FOREST SCENERY. — SLAVE RIVER. — REIN-DEER ISLANDS. — PORTAGES. — NATIVE REMEDIES. — SEPARATE FROM MR. BELL AND HIS PARTY.

IT is probable that the sands of this district and the adjacent limestones belong to the Erie division of the New York system of rocks, considered by the United States geologists to be an upper member of the silurian system, but, by various English naturalists, to be rather part of the Devonian, or of the carboniferous series.

The valley of the Clear-water River, or Washakummow, as it is termed by the Crees, is not excelled, or indeed equalled, by any that I have seen in America for beauty ; and the reader may obtain a correct notion of its general character by turning to an engraving in the narrative of Sir John Franklin's second Overland Journey, executed from a drawing of Sir George Back's. The view from the Cockscomb extends thirty or forty miles, and discloses, in beautiful perspective, a succession of steep, well-wooded ridges, descending on each side from the lofty brows of the valley to the borders of the clear stream which meanders along the bottom. Cliffs of light-coloured sand occasionally show themselves, and near the water limestone rocks are almost every where discoverable. The *Pinus banksiana* occupies most of the dry sandy levels; the white spruce, balsam fir, larch, poplar, and birch are also abundant; and, among the shrubs, the *Amelanchier*, several cherries, the silver-foliaged *Eleagnus argentea*, and rusty-leaved *Hippophäe canadensis* are the most conspicuous.

At the portage, the immediate borders of the stream are formed of alluvial sand; but six or seven miles below, limestone in thin slaty beds crops out on both sides of the river, and, to the left, forms cliffs twenty feet high. A short way further down an isolated pillar of limestone in the same thin layers rises out of the water; and soon after passing it, we come to the White Mud Portage (*Portage de Terre blanche*), of six hundred and seventy paces, where the stream flows over beds of an impure siliceous limestone, in some parts meriting the appellation of a calcareous sandstone, and, for the most part, having a yellowish-grey colour. On the portage, and on the neighbouring islands and flats, the limestone stands up in mural precipices and thin partitions, like the walls of a ruined city ; and the beholder cannot help believing that the rock once formed a barrier at this strait, when the upper part of the river must have been one long lake.   The steep sandy slopes, as they project from the high sides of the valley, appear as if they had not only been sculptured by torrents of melted snow pouring down from the plateau above in more recent times, but that they had been previously subject to the currents and eddies of a lake.   If such was the case, we must admit that other barriers further down were also then or subsequently carried away, as the sides of the valley retain their peculiar forms nearly to the junction of the stream with the Elk River.   I have been informed that the country extending from the high bank of the river towards Athabasca Lake is a wooded, sandy plain, abounding in bison and other game.

In the evening we encamped on the Pine Portage (*Portage des Pins*), which is one thousand paces long.   The name would indicate that the *Pinus resinosa* grows there ; but, if so, I did not observe it, the chief tree near the path being the *Pinus banksiana*, named *Cyprès* by the voyagers. A very dwarf cherry grows at the same place ; it resembles a decumbent willow, and is probably the

Cerasus pumila of Michaux. This is the most northern locality in which it, and the *Hudsonia ericoides*, which was flowering freely at this time, were observed. The *Lonicera parviflora* was also showing a profusion of fragrant, rich, yellow flowers, tinged with red on the ends of the petals, especially before they expand ; and on this day we gathered ripe strawberries for the first time in the season.

July 7th. — The Pine Portage was completed in the morning, and an hour later we crossed the Bigstone Portage of six hundred paces. Afterwards we passed the Nurse Portage (*Portage de Bonne*), of two thousand six hundred and ten paces ; the Cascade Portage, of one thousand three hundred and eighty ; and encamped on the Portage of the Woods, two thousand three hundred and fifty paces long ; where two of our boats were broken. At this place, and on many other parts of the river, smooth granite boulders line the beach. The strata *in situ* are limestone covered by thick beds of sand.

July 8th. — The boats having been repaired early in the morning, we embarked at half-past six, and at eight came to a sulphureous spring, which issues from the limestone on the bank of the river. Its channel is lined with a snow-white incrustation, the taste of the water is moderately saline and sulphureous, and, from its coolness, rather agreeable than otherwise : it had a slight odour of sulphureted hydrogen. Here I obtained specimens of a terebratulite (*T. reticularis*).

In the afternoon we passed the mouth of an affluent named the *Pembina* from the occurrence of the *Viburnum edule* on its banks. I did not observe the fruit of this bush further north than Winipeg River, but I was assured that it grew in various localities up to the Clear-water, beyond which it has not been detected. It is distinguished as a species from the very common cranberry tree, or mooseberry (*Mongsöa meena* of the Crees), by the obtuse sinuses of its leaves; and its fruit has an orange colour, is less acid, more fleshy, and more agreeable to the taste. There is a rapid in Clear-water River just above the Pembina, where a section of the north bank is exposed; and I regretted that I had not leisure to examine it. As seen from the boat in passing, it appeared to be formed of sandstone at the base, then of sand, and high up of shale or sandstone in thin layers. Three miles further down a cliff on the south side, about twenty feet high, is composed of an impure limestone, in very thin layers, capped by a more compact cream-yellow limestone. The sun was intensely hot this day, and, dreading the moschetoes, we avoided the bushy banks of the river, and encamped on an open sand-flat, but did not thereby gain immunity, for we were assailed by myriads during the whole night, a heavy rain having driven them into the tents. The species that now infested us had a light brown colour. Each kind remains in force a fortnight or three weeks, and is succeeded by another more bitter than itself.

The Dog-bane and Indian hemp (*Apocynum androsæmifolium* and *hypericifolium*) grow luxuriantly on the sandy banks of this river. They abound in a milky juice, which, when applied to the skin, produces a troublesome eruption. The voyagers, by lying down incautiously among these plants at night, or walking among them with naked legs, often suffer from the irritation, which resembles flea-bites; hence they designate the plant *herb à la puce*. The second-named species grows more robustly and erectly than the other, and furnishes the natives living on the coast of the Pacific with hemp, out of which they form strong and durable fishing nets.

July 9*th.*—Three miles below our last night's encampment we entered the Elk or Athabasca River, a majestic stream, between a quarter and half a mile wide, with a considerable current, but without rapids.

The lower point of the bank of the Clear-water, where it loses itself in the Elk River, is formed of limestone strata, covered by a thick deposit of bituminous shale, which is probably to be referred to the Marcellus shale of the United States geologists.* The shelving cliff of this shale is one hundred and fifty feet high or upwards, and is capped by sand or diluvium. The high cliffs extend for two or three miles up the Clear-water River, above which the sandy slopes for the most part conceal the strata, except at the water's edge, where the limestone crops out. Much of this limestone has a concretionary structure, and easily breaks down. Other beds are more compact.

The same kind of limestone forms the banks of Elk or Athabasca River for thirty-six miles downwards, to the site of Berens' Fort, now abandoned. The beds vary in structure, the concretionary form rather prevailing, though some layers are more homogeneous, and others are stained with bitumen. The strata for the most part lie evenly, and have a slight dip, but in several places they are undulated, and in one or two localities dislocated, though I did not observe any dykes or intruding masses of trap rock.

Among the organic remains obtained from the beds of limestone at the water's edge, were *Producti,*

* See Appendix for a classification of the rocks of the New York system. The Marcellus shale belongs to the Erie division.

Spirifers, an *Orthis* resembling *resupinata*, *Terebratula reticularis*, and a *Pleurotomaria*, which, in the opinion of Mr. Woodward of the British Museum, who kindly examined the specimens, are characteristic of Devonian strata. In the following season, Mr. Rae picked up from the beach of Clear-water River a fine *Rhynchonella*, which retained chestnut-coloured bands on the shell. The occurrence of colours in fossil shells of so ancient an epoch is very rare. The specimen has been deposited in the British Museum. In one of the cliffs not far below the Clear-water River, the indurated arenaceous beds resting on the limestone contain pretty thick layers of lignite, much impregnated with bitumen, which has been ascertained by Mr. Bowerbank to be of coniferous origin, though he could not determine the genus of the wood.

Fourteen or fifteen miles below the junction of the Clear-water with the Elk River there are copious springs on the right bank. They rise from the summit of an eminence among the fragments of a ruined shale bank, which they have wholly incrusted with tufa. This incrustation, analysed for me by Dr. Fife in 1823, was found to be composed principally of sulphate of lime with a slight admixture of sulphate of magnesia and muriate of soda, and with sulphur and iron. Below this there is a fine section of a bituminous cliff from one hundred and twenty to one hundred and thirty feet high, resting on limestone whose beds are undulated in two directions. The limestone is immediately covered by a thin stratum of a yellowish-white earth, which, from the fineness of its grain, appears at first sight to be a marl or clay. It does not, however, effervesce with acids, is harsh and meagre, and, when examined with the microscope, is seen to be chiefly composed of minute fragments of translucent quartz, with a greyish basis in form of an impalpable powder. This seam follows the undulations of the limestone; but the beds of the superincumbent bituminous shale, or rather of sand charged with slaggy mineral pitch, are horizontal.

About thirty miles below the Clear-water River, the limestone beds are covered by a bituminous deposit upwards of one hundred feet thick, whose lower member is a conglomerate, having an earthy basis much stained with iron and coloured by bitumen. Many small grains and angular fragments of transparent and translucent quartz compose a large part of the conglomerate, which also contains water-worn pebbles of white, green, and otherwise coloured quartz, from a minute size up to that of a hen's egg, or larger. Pieces of greenstone, and nodules of clay-ironstone, also enter into the composition of this rock, which, in some places, is rather friable, in others, possesses much hardness

and tenacity. Some of the beds above this stone
are nearly plastic, from the quantity of mineral
pitch they contain. Roots of living trees and herbaceous plants push themselves deep into beds
highly impregnated with bitumen ; and the forest
where that mineral is most abundant does not
suffer in its growth.

The shale banks are discontinued for a space in
the neighbourhood of Berens' House, where thin
beds of limestone come to the surface, and form
cliffs twenty or thirty feet high at the water's
edge.

Further down the river still, or about three
miles below the Red River, where there was once
a trading establishment, now remembered as *La
vieux Fort de la Rivière Rouge*, a copious spring of
mineral pitch issues from a crevice in a cliff composed of sand and bitumen. It lies a few hundred
yards back from the river in the middle of a thick
wood. Several small birds were found suffocated
in the pitch.

Soon after passing this spot, we saw right ahead,
but on the left bank of the river, a ridge of land
named the " Bark Mountain," looking blue in the
distance, being fully sixty miles off. From its
name, I conclude that the canoe birch abounds on
it. It is the length of a spring day's march, or
about thirty miles, distant from Fort Chepewyan ;

and bison, moose deer, and other game, are said to resort to it in numbers.

At the deserted post named *Pierre au Calumet*, cream-coloured and white limestone cliffs are covered by thick beds of bituminous sand. Below this there is a bituminous cliff, in the middle of which lies a thick bed of the same white earth which I had seen higher up the river in contact with the limestone, and following the undulations of its surface.

A few miles further on, the cliffs for some distance are sandy, and the different beds contain variable quantities of bitumen. Some of the lower layers were so full of that mineral as to soften in the hand, while the upper strata, containing less, were so cemented by iron as to form a firm dark-brown sandstone of much hardness. The cliff is, in most places, capped by sand containing boulders of limestone. One very bituminous bed, carefully examined with the microscope, was found to consist, in addition to the bitumen, of small grains of transparent quartz, unmixed with other rock, but enclosing a few minute fragments of the pearly lining of a shell. A similar bed in another locality contained, besides the quartz, many scales of mica. The whole country for many miles is so full of bitumen that it flows readily into a pit dug a few feet below the surface.

In no place did I observe the limestone alternating with these sandy bituminous beds, but in several localities it is itself highly bituminous, contains shells filled with that mineral, and when struck yields the odour of *Stinkstein*. It is probable that the whole belongs to the same formation, but I do not possess evidence of the facts to satisfy a geologist.

The rate of our descent of the Elk River must this day have exceeded six geographical miles an hour, indicating a strong current. This river, named also the Athabasca or *Rivière la Biche*, rises in the parallel of $47\frac{1}{2}°$ north latitude, near the foot of Mount Brown, a peak of the Rocky Mountains, having a height of sixteen thousand feet above the sea. Its course in a straight line to the influx of Clear-water River is three hundred miles ; but the river course, including its windings, must be more than one third greater. The elevation of its sources is probably seven or eight thousand feet. Lesser Slave Lake, situated about midway between its origin and the junction above mentioned, lies, according to Captain Lefroy, eighteen hundred feet above the sea. Some of the feeders of the Oregon spring from very near the head of the Athabasca, and many tributaries of the Saskatchewan arise not far to the southward. It is the most southern branch of the Mackenzie ; and as it originates further from the mouth of that great river than any other affluent, it may be considered as its source. It flows partly through prairie lands, and its Canadian appellation of *Rivière la Biche* indicates that the American red-deer, or Wapiti, frequents its banks. Its English name of Elk River, having reference to the moose deer, is a mistranslation of the Canadian one, and is also inappropriate as a distinctive epithet, though the moose grazes on its banks, as well as on the Mackenzie, down to the sea. The *Wapiti* is not known on Slave River or Lake, but further to the west it ranges as far north as the east branch of the River of the Mountains near the 59th parallel, where Mr. Murdoch M'Pherson informs me that he has partaken of its flesh. From the Saskatchewan and Lesser Slave Lake the country can be traversed by horsemen who are sufficiently acquainted with the district to avoid the deep ravines through which the streams flow. By this route a band of horses were brought to Methy Portage in August, 1848, though they were too much exhausted by their journey to be of service. In 1849 a fine body of upwards of forty horses came to the portage from Lesser Slave Lake, early in the season and in good condition.

July 10*th.* —— Our voyage this morning was impeded by a strong head-wind, followed by heavy rain, which compelled us to put ashore for four or five hours. We were able to resume our route at 10 A.M., and at noon we came to high sandy banks named *Les Ecores*, resembling the sandy deposits on the Clear-water River. These continue down to the alluvial delta formed by the four or five branches into which the river splits before entering the Athabasca Lake, or Lake of the Hills.

At 5 P.M. we arrived at the head of this delta, and, passing down the main channel, held on our way till 8 o'clock, when we landed to cook supper, and then re-embarked to drift with the current during the night, the crews, with the exception of the steersmen, going to sleep in the boats.

July 11*th.*—We entered Athabasca Lake at three in the morning, but found, to our mortification, that two of the boats, through the inattention of the steersmen, had taken a more easterly branch of the river in the night, which would delay their arrival at Fort Chepewyan for some hours, and consequently be the means of detaining us for that time.

Immediately on emerging from the river we saw the Mackenzie River brigade of boats crossing the lake towards the entrance of the Embarras River, lying four or five miles to the westward of the branch we had descended. On our firing guns and hoisting the sails and ensigns, we were per-

ceived by the officer in charge of the brigade,
Chief Factor Murdoch M'Pherson, who waited till
we joined him.   From this gentleman I received
much useful intelligence of the measures he had
taken for supplying the expedition with provisions
during our winter residence in Fort Confidence, at
the north end of Great Bear Lake, and also a list
of all the provisions and stores remaining at Fort
Simpson, the Company's chief post and depôt on
the Mackenzie ; and I have pleasure in acknow-
ledging here, that I am indebted to him for much
invaluable assistance, as well as for very many
acts of personal kindness.

To him we committed the last letters that we
could send to our families and friends in Europe
this year.   I had sent despatches to the Admiralty
from Methy Portage, not being sure that we should
meet the Mackenzie River brigade, which is the
latest that goes out.   It can seldom cross Great
Slave Lake before the end of June, and from
twenty to twenty-four days are required for the
passage of loaded boats from thence to Methy
Portage.   There the Mackenzie River party are
met by a brigade from York Factory, which brings
up goods for next year's supply of the northern
posts, and takes back the furs brought from
the Mackenzie.   There is just time in common
seasons for that brigade to descend to York Fac-

tory before the annual ship sails from thence for England, about the middle of September, or in backward seasons a week or two later; and afterwards to return to the colony at Red River, where the crews reside, and from whence they come annually in the spring on this special service. For many years the Methy Portage brigade has been conducted by a guide named L'Esperance, and on that account it is known by the name of L'Esperance's brigade.

After the return of the Mackenzie River boats to Fort Simpson, the winter's supply of goods has to be sent to the outposts; but as some of these are at the distance of four or five weeks' travelling, the parties carrying them are not unfrequently arrested by frost, far from their destination, and the posts suffer severely,—sometimes to the length of actual starvation and loss of life; an instance of which occurred before I left the country.

We reached Fort Chepewyan at half-past 7 A.M., but the two boats that strayed from us did not arrive till the afternoon, and the chief artizans being in the missing boats, the intention I had of giving them a complete repair here, and putting on false keels, was frustrated. Their leaks were, however, stopped, and some planks replaced, which detained us till 11 A.M. on July the 12th, when we left the fort.

The height of Lake Athabasca above the sea is estimated by Captain Lefroy at six hundred feet.[*] Its basin offers another instance of the softer strata having been swept away at the line of their junction with the primitive rocks; and a reference to the map will show that there must have been an evident connection between the cause of this excavation and that of Wollaston and Deer's Lakes, belonging to the Missinipi River system.[†] Wollaston Lake is said to supply a river at one end, which falls into Athabasca Lake, and one at the other, which joins the Missinipi, which, if correct, is not a common occurrence in hydrography, though one or two instances of the kind, in seasons of flood, have been alluded to in the preceding pages.

Much of the country in the immediate vicinity of Chepewyan is composed of rounded knolls of granite, nearly destitute of soil, and many of them smooth and polished. These rocks extend along the north shore of the lake; and the eminences rise in the interior in a confused manner, one over

* Eight months of observations with the boiling-water thermometer by this officer, give an elevation of 468 feet, excluding two observations on which he could not rely. This being, however, in his opinion too low, he assigns the altitude mentioned in the text, after a review of his entire body of observations in various parts of the country, and checking one by another.

† See Appendix.

the other, to the height of four or five or perhaps six hundred feet above the water. They also form many islands at the west end of the lake and in front of the fort. Between this end of the lake and the mouth of Peace River there lies a muddy expanse of water, named Lake Mamawee; and, in times of flood, the waters of Peace River flow by this channel into Athabasca Lake, rendering its usually transparent waters very turbid. A short way to the eastward of the fort a grey gneiss rock is associated with reddish granite, and its beds are much contorted and are traversed by veins of vitreous quartz. Still further off in that direction a cliff of chlorite-slate occurs. Plumbago of excellent quality has been found on the shores of this lake, and I have been informed that at its eastern extremity, named the *Fon du Lac*, there is much sandstone — the resemblance to the succession of strata on Lake Superior being maintained here also. Granite rocks, generally forming rounded knolls, prevail in Stony River, by which name the discharge of Athabasca Lake is known, and on whose banks we encamped on the evening of the 12th.

Soon after starting on the morning of the 13th, we passed the mouth of the Peace River, or Unjugah, which is the largest branch of the Mackenzie, since it brings down more water than either

K 3

the Athabasca River or River of the Mountains. When it is flooded it overcomes the stream of Stony River, and carries its muddy waters into Lake Athabasca, meeting there another rush of waters coming through Lake Mamawee; but at other times there is a strong current in Stony River, and at one point a dangerous rapid, where a gentleman of the North-west Company was drowned many years ago. A delta, intersected by several channels, exists at the junction of Peace River with Athabasca Lake and its outlet. The source of Finlay's branch of this river is nearly in the same parallel with its mouth, but in its course the trunk of the river makes a great curve to the southward, and its southern tributaries rise in the same mountains from which Frazer River issues on the west side of the Rocky Mountains, the upper waters of the Peace River coming in fact through a gap in the chain which forms one of the passes leading to the Pacific coast. Captain Lefroy, who has travelled through this district, makes the following remarks upon its elevation. " The next series of observations was made in the elevated region at the base of the Rocky Mountains, between Peace River and the Saskatchewan, a district remarkable for its gradual and regular ascent, preserving throughout much of the character of a plain country. From Lake Athabasca

to Dunvegan, a distance of about six hundred and
fifty miles " (250 geographical miles in a straight
line), " there occurs but one inconsiderable fall, and
a few rapids ; the bed of the Peace River preserves a
nearly uniform inclination, in which it rises three
hundred and ten feet. The stream is, however, more
rapid above Fort Vermilion than below it. The
depth of the bed of the river below the surrounding country increases with great uniformity as we
ascend the river. A defile, very similar to that called
the Ramparts on Mackenzie's River, but on a finer
scale and with far more picturesque features, occurs
about eight miles above the river Cadotte, in long.
117°; and here the river has cut a passage through
cliffs of alternating sandstone and limestone to a
bed of shale, through which it flows, at a depth
of two hundred feet (by estimation) below their
summit. The general elevation of the country,
however, still continues to increase, and at Dunvegan, it is six hundred feet above the bed of the
stream ; yet even at this point, except on approaching the deep gorges through which the tributaries
of Peace River join its waters, there is little indication of an elevated country ; the Rocky Mountains
are not visible, and no range of hills meets the eye.
A rough trigonometrical measurement gave five
hundred and thirty-eight feet as the elevation of
Gros Cap, a bold hill behind Dunvegan, above the

bed of the river; and the ground was estimated
to rise behind Gros Cap, by a gradual ascent, until
it attains the general level." (*Lefroy*, l. c.)   The
elevation above the sea, that this intelligent officer
assigns to the country about Dunvegan is sixteen
hundred feet, and the region in which the sources
of the river occur is probably four times as high.

The oaks, the elms, the ashes, the Weymouth
pine, and pitch pine, which reach the Saskatchewan basin, are wanting here, and the balsam-
fir is rare; but as these trees form no prominent
feature of the landscape in the former quarter, no
marked change in the woodland scenery takes place
in any part of the Mackenzie River district until
we approach the shores of the Arctic Sea.   The
white spruce continues to be the predominating
tree in dry soils whether rich or poor; the Banksian pine occupies a few sandy spots; the black
spruce skirts the marshes; and the balsam-poplar
and aspen fringe the streams; the latter also springs
up in places where the white spruce has been destroyed by fire.   The canoe-birch becomes less
abundant, is found chiefly in rocky districts, and is
very scarce north of the arctic circle.   It still,
however, attains a good size in the sheltered valleys
of the Rocky Mountains, up to the 65th parallel.
Willows, dwarf birches, alders, roses, brambles,
gooseberries, white cornel, and mooseberry, form

the underwood on the margins of the forest; but there is no substitute for the heath, gorse, and broom, which render the English wild grounds so gay. On the barren lands, indeed, the heath has representatives in the Lapland rhododendron, the *Azalea*, *Kalmia*, and *Andromeda tetragona*, but these are almost buried among the *Corniculariæ* and *Cetraria nivalis* of the drier spots, or the *Cetraria islandica* and mosses of the moister places, and scarcely enrich the colours of the distant hills.

The granite knolls show themselves at frequent intervals on the banks of Slave River, which is the appellation of the stream formed by the junction of the Peace and Stony Rivers; and in several places, ledges of the rock crossing the river form rapids. One of these is named the Lightening Place of the Hummock, because it occurs at the beginning of a reach two miles long, which is terminated by a sandy bluff on the right bank, twenty or thirty feet high, and covered with Banksian pine. This *Bute*, as it is termed by the Canadian voyagers, is about thirty miles from Fort Chepewyan, and opposite to it there is a limestone cliff, constructed of thin undulated layers. The lower beds of the limestone have a compact structure, a flat conchoidal fracture, and a yellowish grey colour. Some of the upper beds contain mineral pitch in fissures, and shells, which Mr.

Sowerby in 1827 ascertained to be *Spirifer acuta*, and several new *Terebratulæ*, one of them resembling *T. resupinata*; associated with them a *Cirrus* and some crinoidal remains occur. Not far above this cliff, a vitreous reddish-coloured sienite protrudes; and half a mile or so below it, the stream passes between rounded hummocks of granite, one of which forms an island, the water-course evidently following the line of dislocation of the strata. The clustered nests of large colonies of the republican swallow (*Hirundo fulva*) adhere to the ledges of the limestone cliffs, and the bank swallow (*Hirundo riparia*) has pierced innumerable holes in the sandy brows.

A small tributary enters the river from the left, behind an island, lying a short way below the *Bute*, and another comes in from the right, beneath which the brown vitreous sienite re-appears, forming a flatly rounded eminence. Within a mile of this pyrogenous rock, another limestone cliff occurs on the left bank, at the commencement of a pathway which leads over prairie-lands, or through sprucefir woods, marshes, and by small lakes, to the Salt River, to be hereafter noticed.

A mile and a half below this are the three Rocky Islands (*Isles des Pierres*), which is perhaps the best locality on the river for studying the connection of the limestone with the pyro-

genous rocks; and I regretted that I could devote
no time to this purpose.   The beds of limestone,
as seen in passing rapidly along these islands,
appeared of various thickness, some being thin
and shaly, and almost all more or less undulated,
saddle-formed, or contorted.   On the borders of a
channel between two of the islands, a conglomerate
is interleaved with sienite; and in the vicinity there
are beds of a brownish, finely crystalline limestone,
having a conchoidal fracture, the fragments being
sharply angular.   The conglomerate varies considerably in its texture in different layers, and
even in different parts of the same bed.   It contains, in general, a large proportion of small
rounded grains of translucent or milky quartz,
with angular fragments of various sizes of vitreous
quartz, chlorite-slate, and calc-spar, imbedded in a
powdery or friable white basis, which does not
effervesce with acids; the whole forming a tough
stone.   In some beds the quartz grains predominate, so as to render the rock a coarse sandstone;
but in other parts, these grains appear to have
been fused into a bluish quartz rock, the original
granular structure being only faintly discernible,
and to be detected chiefly in spots, where some of
the powdery basis remains unchanged.   In one
bed, angular fragments of greenstone encrusted
with calc-spar occur.   The sienite contains grains

of hornblende and quartz in about equal quantities, imbedded in a snow-white powdery basis, which appears to be disintegrated felspar.

A mile below the Stony Islands we passed the smaller Balsam Fir Island, below which there is a pretty little *bute* on the left, where the purplish-coloured rock that protruded appeared to us in passing to be amygdaloid or porphyritic trap rock. Some miles further down we entered among the rather high and rocky cluster of the Rein-deer Islands (*Isles de Carrèbœuf*) by a channel having a north-north-west direction. The rocks here appeared to us as we shot past them to be principally trap, associated with gneiss, or perhaps chlorite-slate. A point on the main shore, on which I landed in 1820, is composed of felspar and quartz, and is probably a variety of granite.

A short way further down the Great Balsam Fir Island (*La grand Isle des Epinettes*), which is a mile across and three or four long, has a triangular form, and divides the river into two channels. We descended the easternmost, or right-hand one, which is the most direct, and has a high and sandy eastern bank.

Below this a bend of the river is filled with many rocky islands, occasioning numerous rapids and cascades, and seven or eight portages. The river expands here to the width of a mile and a half or two miles; its bed is every where rocky, and the rocks are apparently all primitive; but as the boat-route lies wholly through the eastern channels, we had no opportunity of inspecting the opposite shore closely. The islands are well wooded, and the scenery picturesque. Some of the narrower channels, which would be convenient for the descent of boats, are blocked up by immense rafts of drift timber, which have been accumulating for many years, and which could not be set free without very great and long-continued labour. Large flocks of pelicans have made their nests on the more inaccessible rocks rising from the brows of the cascades. In the evening we ran down the Dog Rapid after lightening the boats, and afterwards descended a second rapid, and then encamped on a smooth granite rock early in the evening, there not being time to complete the Chest Portage before dark.

Embarking at 3 A.M. on the 22d of July, we descended a narrow channel to the Chest Portage (*Portage de Cassette*), where our five boats were hauled over a pathway of four hundred and sixty-five paces, and their cargoes carried. A rocky chasm at this place, being one of the numerous channels through which the water flows, encloses a perpendicular cascade upwards of twenty feet high; beneath which an isolated column of rock divides the current into two branches, which eddy with great force into the niches and recesses of the stony walls. Huge angular blocks obstruct the water-course, and drift trees, entangled among them, partially denuded of their branches, and wholly of their bark, point in all directions. The over-hanging woods almost seclude this gloomy ravine from the sun; and it presents such an aspect of wildness and ruin as rarely occurs even in this country. In one part of the portage road a bed of gneiss is flanked on each side by masses of granite. A labyrinth of passages among granite rocks exists below the portage, many of them en-tirely choked up with drift timber. In passing rapidly through one of them we grazed a point composed of a crumbling red and grey porphyritic rock, perhaps an amygdaloid; many cubical and irregularly angular fragments had fallen from it.

At the Island Portage, which immediately fol-lowed, the cargo is carried only in the ascent of the river. Our boats descended the fall with their entire load. We next crossed the Raft Portage (*Portage d'Embarras*), which occupied us three hours. At the Little Rock Portage, which follows, the rock is composed of felspar, quartz, and chlorite, being the *protogine* of Jurine. It differs from the slaty rock observed near the Rein-deer Islands, in not being stratified. At the Burnt Portage, the next in order, the rock, which is a porphyritic granite, acquires a polished glistening surface. There is a cascade here of fifteen or twenty feet. The succeeding portage, named the " Mountain," from the steep bank down which the boats are lowered, is shorter than the others, being only one hundred and seventy paces across. The rock at this place is a red, compact, shining or vitreous-looking granitic porphyry, much fissured, and breaking, by the action of the frost, into cubical or rhomboidal blocks, sometimes of great size. The principal fissures are generally, but not always, parallel to each other, and may be traced for seventy or eighty yards without a break, in a transverse direction to that of the eminences and projecting tongues of the rock. Their course is north-east by north, and south-west by south ; and they are, for the most part, four or five feet apart. The minor cracks meet the chief ones at various oblique angles, and sometimes cross them, but not generally. At another denuded point of rock, the wider cracks crossed each other, one set running east-south-east and west-north-west. The recesses left by the blocks which fall away retain their sharp-cornered rectangular shape. A layer of hornblende-slate or basalt shows itself at one spot.

The launching-place for the boats here is both steep and rugged ; and a brigade seldom passes without some of the boats being broken. One of ours was injured; but, being soon repaired, we left the portage by six in the evening, and encamped for the night at the south end of the Pelican Portage, which is seven hundred paces long.

The power of the sun, this day, in a cloudless sky, was so great, that Mr. Rae and I were glad to take shelter in the water while the crews were engaged on the portages. The irritability of the human frame is either greater in these northern latitudes, or the sun, notwithstanding its obliquity, acts more powerfully upon it than near the equator; for I have never felt its direct rays so oppressive within the tropics as I have experienced them to be on some occasions in the high latitudes. The luxury of bathing at such times is not without alloy; for, if you choose the mid-day, you are assailed in the water by the *Tabani*, who draw blood in an instant with their formidable lancets; and if you select the morning or evening, then clouds of thirsty moschetoes, hovering around, fasten on the first part that emerges. Leeches also infest the still waters, and are prompt in their aggressions.

The *Geum strictum* grows plentifully on these portages, and is used by the natives for the purpose of increasing the growth of their hair. They dry the flowers in the sun, powder them, and mix them with bear's grease. The *Eleagnus argentea*, which is also abundant on the banks, is named by the Chepewyans *Tâp-pah*, or grey berry. It is the bear-berry of the Crees, and the stinking willow of the traders ; so called, because its bark has a disagreeable smell.

July 15*th*. — The portage was completed, and breakfast prepared and eaten, in five hours and a half. At the lower end of the path, a sienitic rock, composed of crystallised quartz, aurora-red felspar, and greenish-black hornblende, yields large cubical blocks of a handsome stone. One of the small boats was overset in lowering it down a narrow channel, and the oars, a coil of rope, and the boat-lockers were swept away by the current. A boat's anchor, and some clothes belonging to two of the crew, were in one of the lockers.

An hour before noon we had crossed the Portage of the Drowned (*Portage des Noyes*), where granite is the prevailing rock. This being the last of the portages, three of the small boats brought from England were stowed with pemican for the sea-voyage ; and Mr. Bell was left to follow with the large boat and the fourth small boat, containing the stores for house-building, nets, ammunition, and other supplies for winter use. He would have accompanied us ; but his men had to make oars in place of those which had been lost ; an employment which was likely to occupy them for two or three hours.

CHAP. V.

PYROGENOUS ROCKS. — RATE OF THE CURRENT OF SLAVE RIVER. — SALT RIVER AND SPRINGS. — GEESE. — GREAT SLAVE LAKE. — DOMESTIC CATTLE. — DEADMAN'S ISLANDS. — HORN MOUNTAIN. — HAY RIVER. — ALLUVIAL LIGNITE BEDS. — MACKENZIE'S RIVER. — MARCELLUS SHALE. — FORT SIMPSON. — RIVER OF THE MOUNTAINS. — ROCKY MOUNTAINS. — SPURS. — ANIMALS. — AFFLUENTS OF THE MACKENZIE. — CHETA-UT-TINNE.

No primitive rocks were seen on the route down the Mackenzie, on this voyage, after leaving the Portage of the Drowned; but in 1820, when we crossed Great Slave Lake, near the 113th meridian, we traced the western boundary of these rocks, from near the mouth of Slave River, northwards by the Rein-deer Islands to the north side of the lake, and continued to travel within their limits up to Point Lake in the 66th parallel. The western edge of the formation was afterwards found at the north-east and eastern arms of Great Bear Lake.

The district intervening between the granite at the Portage of the Drowned and the Salt River is flat, with sandy terraces and slopes rising from the river to the height of from twenty to eighty feet, there being in some places two in others three or more such terraces, while in others the river has made a section of the sandy deposit, and formed a high and steep cliff. The valley of the river, deflected to the westward by the rocks of the portages, passes here through the more level (upper ?) silurian strata.

At Gravel Point (*Pointe de Gravoir*), ten miles from the portages, a bed of concretionary or brecciated limestone protrudes from under a sand-bank forty feet high, and two miles higher up a cliff of cream-coloured and brownish limestone stands on the right bank. The country on both sides of the river there appears to be a plain, which has a general level of about fifty feet above the bed of the stream.

Just before arriving off the mouth of Salt River, we picked up one of the boat's lockers containing the anchor, which had been carried away fifteen miles higher up nearly eight hours before, so that it drifted about two miles an hour, including the time it might have been detained in eddies.

In 1820, I ascended the very tortuous Salt River, for twenty miles, for the purpose of visiting the salt springs, which give it its name. Seven or eight copious springs issue from the base of a long even ridge, some hundreds of feet high, and, spreading their waters over a clayey plain, deposit much pure common salt in large cubical crystals. The *mother water*, flowing off in small rivulets into the Salt River, communicates to it a very bitter taste; but before the united streams join the Slave River, the accession of various fresh-water rivulets dilutes the water so much that it remains only slightly brackish. A few slabs of greyish compact gypsum protrude from the side of the ridge above mentioned; and a pure white gypsum is said to be found at Peace Point on Peace River, distant about sixty or seventy miles in a south-south-west direction, whence we may conjecture that this formation extends so far. From the circumstance that the few fossils gathered from the limestone on Slave River are silurian, I venture to conjecture that these springs may belong to the Onondago salt group of the Helderberg division of the New York system. The Athabasca and Mackenzie River districts are supplied from hence with abundance of good salt. We obtained some bags of this useful article from Beaulieu, who was guide and hunter to Sir John Franklin on his second overland journey, and who has built a house at the mouth of Salt River. This is a well chosen locality for his residence: his sons procure abundance of deer and bison meat on the salt plains, which these animals frequent in numbers, from their predilection for that mineral; and Slave River yields plenty of good fish at certain seasons. It is the most southern locality to which the *Inconnu*

or *Salmo Mackenzii* comes, on this side of the
Rocky Mountains, as it is unable to ascend the cascades in the Slave River. The *Coregoni* are the
staple fish of the lakes here, as they are elsewhere throughout the country ; and there are also
pike, burbot, and excellent trout. A limestone cave
in the neighbourhood, which was too distant for us
to visit, supplies Beaulieu with ice all the summer,
and he gave us a lump to cool water for drinking,
which was extremely grateful. The ammunition
and tobacco with which I repaid these civilities were
no less acceptable to him. Indeed, I believe that
he turns his residence on the boat-route to good
account, as few parties pass without giving him
a call.

After a short halt, we resumed our voyage until
7 P.M., when we landed to cook supper, after
which we re-embarked to eat it ; and, having
lashed the boats together, drifted down the stream
all night, one man being appointed to steer.

July 16*th.* — Though we lay down in the best
manner we could in the boats during the night,
the continuous assaults of the moschetoes deprived
every one of rest, and rendered us all so feverish,
that we were glad when daybreak called the
crews to the oars, and the boats acquired motion
through the water, by which we obtained some
relief.

The sandy banks of the river show sections in many places upwards of twenty feet high ; and, in almost all, the sand is distinctly stratified ; the layers being of different colours, and often having clayey or loamy seams interposed. The whole of the banks, from Salt River downwards to Slave Lake, appear to be alluvial ; and many small lakes existing behind them communicate with the river by narrow channels. In ordinary seasons at this date, vast numbers of Canada geese moult in the district, and are followed by their young brood not yet fully fledged, which fall a ready prey to the natives or voyagers descending the river. In 1825 I could have filled a boat with these delicate young birds. This year, owing to the high waters, the greater part of the broods had retreated to the lakes, where grass could be more easily procured, and we obtained only a few. The natives observe, that, besides the old birds which rear young, and moult when their offspring are obtaining their plumage, there are a considerable number who do not breed, but keep in small bands, and are called " barren geese." Of these we saw some flocks ; but they were not easily approached without a greater loss of time than we could spare.

We kept at the oars all day, except when we landed to breakfast, or to cook supper, and, after sunset, resumed the plan of drifting, with very little

better success, as far as sleep was concerned, than on
the preceding night. During the day the sun's rays
felt intensely warm ; and the puffs of northerly
wind blew as hot as if they had passed over the
deserts of Arabia.   At midnight a strong contrary
wind springing up, compelled us to anchor until
half-past 2 A.M. on the 17th, when we again took
to the oars, and entered Great Slave Lake at 7
in the morning.

Like the Athabasca River, the Slave River joins
its lake through a delta of low, well-wooded, alluvial
islands, by many channels, having a spread of more
than twenty miles. Near the easternmost, which is
named John's River (*Rivière à Jean*), is Stony
Island, a naked mass of granite, rising fifty or
sixty feet above the water ; and beyond that, to
the eastward, the banks of the lake are wholly
primitive.   In the vicinity of the westernmost
channel of the delta, and from thence to the efflux
of the Mackenzie, the whole southern shore of the
lake is limestone, associated with a bituminous
shale, and belonging, as well as can be ascertained
from its fossils, to the Erie division of the New
York system, which includes the Marcellus shales,
and is referred by English geologists to the carboniferous series.   In the small channel which divides
Moose-deer Island from the point of the bay on
which the present Fort Resolution stands, many

boulders of porphyritic and common granites, greenstone, and limestone occur; also large angular blocks, not worn or rounded, of a conglomerate of granite, chert, and hornblende rocks cemented by a basis of ironstone.

We reached Fort Resolution at 10 A.M.; and having received some supplies of fish, and two or three deals for repairing the boats, we resumed our voyage, after a halt at the fort of one hour. Domestic cattle have been introduced at this place, and at the posts generally throughout the country, even up to Peel's River and Fort Good Hope, within the arctic circle. At this season the moschetoes prevent them from feeding, except when urged by extreme hunger; and fires are made for their accommodation near the forts, to which they crowd, and, lying to leeward amidst the smoke, ruminate at their ease. Smoke is the only remedy against these venomous insects; and at this time of the year, when the heat renders a free circulation of air in the houses essential, the rooms are made comfortable by nailing bunting over the windows, and burning turf or rotten wood in a pan on the threshold of the door. At no place on our route were the moschetoes in denser clouds than this day at Fort Resolution; and we gladly left them behind as we launched into the lake with a favourable breeze. We had not gone above two miles, when we saw Mr. Bell and his boats issuing from behind Moose-deer Island, and steering for the house; but time was too precious for us to wait for his coming up. Our route lay through a small group of islands lying five or six miles off the bay into which Buffalo Creek falls. In these islands, a bituminous limestone crops out in thin horizontal layers near the water's edge; but, except in a few places, its beds are concealed under a beach composed of fragments of the same stone, partly rolled and worn, partly with recently broken edges. The islands are most of them low; and the stony beach rising above their centres encloses marshy spots traversed by ridges of sand and gravel, more or less wooded.

At 5 P. M. the wind, which had been increasing all the afternoon, rose to a high gale, and we put into a good boat harbour at Deadman's Island and encamped. This spot received its name from a massacre committed by a war party of Beaver Indians, who surprised a body of Dog-ribs encamped there, and destroyed them all. Thirty years ago many of the bones of the victims were to be seen, but they have now disappeared. The influence of the Hudson's Bay Company has put a stop to these war excursions, and tribes formerly the most hostile to each other now meet in amity at the trading posts.

This lake is a breeding station of the *Sterna cayana*. The arctic tern also hatches on its shores, depositing its eggs among the gravel on the beach. The leaves of the gooseberry bushes had been stripped off by a black-banded caterpillar, and it was evident that the fruit would fail this season. The ice having parted from the shore little more than a fortnight, vegetation was backward. A strong gale, bringing on a keen frost, blew all night, and effectually quelled the moschetoes, so that, though we could not but regret the detention, we all enjoyed some hours of sound repose. But in the morning of July 17th, during one of the squalls of a thunderstorm accompanied by heavy rain, the tent pegs drew from the sandy soil on which we had encamped, and the dripping canvass falling upon us put an end to our rest. We were miserably cold and wet before this mischance was remedied. High winds and a rolling sea kept us stationary all day, and our carpenters took advantage of the delay to secure the thwarts of two of the boats which had given way on the portages.

July 18th.——The gale did not abate so as to allow us to embark until 4 P.M., when we resumed our voyage, and at 9 encamped again in a small boat harbour at Burnt Point. This coast of the lake generally is flat and shelving, and secure landing-places for boats are very scarce. Though we did not discover limestone *in situ* here, the beach is formed of fragments of that stone of very various size, mixed with some bituminous shale, and a few granite boulders. This point is about thirty-five geographical miles from Fort Resolution. In a bay a little to the westward several sulphureous streams issue from a limestone containing corals. The channels of these streams are encrusted with a similar tufa to that observed on Clear-water and Athabasca Rivers, and the organic remains that have been examined indicate the formations to be of the same geological epoch.*

July 19th.—Embarking at three, we passed the mouth of Buffalo Lake River, and after five

---

* The fragments of black and bituminous shale which strew the beaches of these islands, and which evidently have not travelled far, contain a "pteropodous shell (*Theca*) apparently the *Tentaculites fissurella* of Hall, a *Chonetes*, the *Strophenema setigera* of Hall, and *Avicula lævis* of the same author; at least they are undistinguishable from his figures of these fossils in the Marcellus shale, which according to him is upper silurian, but is probably somewhat newer, and what we call Devonian. Two corals in the associated bituminous limestone are characteristic of the same epoch, namely a *Strombodes* of Hall, having its cysts filled with bitumen, and a *Favosites* very like the common *F. polymorpha* of the Plymouth marbles. I have not identified any of the *Terebratulæ* from Great Slave Lake, but they are certainly either Devonian or carboniferous, and not silurian. There is nothing like a secondary fossil in the collection."—*Woodward* in lit.

hours' pulling, put into Canoe or Sandy River to cook breakfast. From this place a rising ground on the north side of the lake is distinctly visible, the distance being about thirty miles or more. The Horn Mountain, an even ridge more to the westward, appeared also in the extreme distance, being at least sixty miles off.

Hay River enters the lake at the distance of eleven or twelve miles from Canoe River. It is formed of two branches, the westernmost of which rises from Hay Lake and the other one originates not far from the banks of Peace River, and flows past Fort Vermilion. Hay River Fort, now abandoned, stood at the junction of the two. On the eastern branch, the country is an agreeable mixture of prairie and woodland, and this is the limit of those vast prairies which extend from New Mexico. Below the forks of Hay River the country is covered with a forest intersected by swamps.

The range of the Wapiti is nearly coincident with the boundaries of these prairies. The bison, though inhabiting the prairies in vast bands, frequents also the wooded country, and once, I believe, almost all parts of it down to the coasts of the Atlantic; but it had not until lately crossed the Rocky Mountain range, nor is it now known on the Pacific slope, except in a very few places. Its most northern limit is the Horn Mountain mentioned above. The musk-ox does not come to the south of the arctic circle.

In the evening we pitched our tents on a small island, being one of a group known as Desmarais's Fishery, and from a party of Indians encamped on a neighbouring point we obtained a supply of fish.

The whole south shore of the lake, westward from Slave River, is low and level, and is lined in parts, and especially between Hay River and Desmarais's Islands, by rafts of drift-timber, which, pressed on shore by northerly winds, become waterlogged and covered with sand mixed with comminuted wood and decaying grasses. As this buried forest accumulates, willows and balsam poplars spring up from its surface, and bind all together by their roots. The swamps that extend backwards from the lake appear to have originated very much in this way; and had the locality been one where there was much drift-sand, the erect trees might in parts have been swallowed up and killed by sand-drifts, having their roots in the subjacent lignite or shale. In Slave Lake, however, the sand does not act so conspicuous a part as in Lake Winipeg. The recent deposits conceal the limestone strata, except at a very few places, but the numerous fragments which line the beach show that a bituminous limestone, associated with a black shale having a resinous streak, and a thin marly slate, must exist in the neighbourhood. They are referrible, as has been mentioned above, to the Marcellus shale. At the Stony Point, between Hay River and Desmarais's Fishery, fragments of these rocks form the beach, on which some very large boulders of gneiss, sienite, and greenstone also lie. At Desmarais's Fishery I observed the same kind of beach, with the addition of blocks of basalt, of a dull-red sandstone, a coarse conglomerate composed of rounded pieces of sandstone cemented by a basis of red clay strongly impregnated with iron. The limestone fragments contained bivalves and corals. On the 6th of July, in the following year, the whole bay that we had traversed in this day's voyage was filled with ice, not yet parted from the shore ; and the lake is scarcely ever navigable in this quarter before the beginning of the month *, so that we were only a fortnight later than we could have hoped to cross the lake, had the boats advanced even to Isle à la Crosse the first season, as they might have done under a very favourable combination of circumstances. But this fortnight, by enabling the expedition to be at the mouth of the Mackenzie on the disruption of the ice of the Arctic Sea, would have been of the very greatest

* Dean and Simpson made their way through it on the 24th of June ; the ice, however, was still adhering to the shore at some points.

advantage.   In fact, a fortnight is no contemptible
portion of the six weeks during which the Arctic
Sea is navigable for boats.   The ice on this lake
is sometimes eleven feet thick ; at Fort Resolution,
and at Big Island, which lies across the western
oulet of the lake, it varies from five to seven feet.

July 20*th.*——This morning we crossed from Desmarais's Fishery obliquely to the north side of the
lake, through an archipelago of islets and along the
south side of Big Island.   There is more or less
current in the passages, and from the general
shallowness of the water, it is probable that the
limestone strata come near the surface, but they are
concealed by gravel and boulders.   To the south
of this traverse, on a strait two miles wide, which
separates the site from Big Island, stood formerly
Fort George.   The limestone beds are said to crop
out in its neighbourhood.

During the whole summer, in the eddies between
the islands of this part of the lake, multitudes of fish
may be taken with hooks and by nets, such as trout,
white fish, pike, sucking-carp, and *inconnu.*   In
spring and autumn wild-fowl may be procured in
abundance at several places in the neighbourhood,
which are their accustomed passes ; and the fishery
on the north side of Big Island seems to be inex-
haustible in the winter.   With good fishermen and
a proper supply of nets, a large body of men may be

wintered here in safety and plenty, and it was to this place that I contemplated conducting any of the crews of the Discovery ships that we might be so happy as to find. To it, also, I purposed to send large portion of my party in the winter of 1848–9. In no other part of the Hudson's Bay Company's territories, that I am acquainted with, can so many people be maintained, with so much certainty, on the resources of the country. A body of good native hunters, well supplied with ammunition, could not fail to bring from the Horn Mountain an agreeable variety of diet in form of rein-deer and bison meat, and in some seasons the American hare may be snared in great numbers.

After we had rowed about thirty-four or thirty-five miles from our encampment of the preceding night, the funnel-shaped entrance of the river had contracted to a width of about two miles, and the current, as it washed the boulders of the beach, made a bubbling noise, like that of a strong rapid; and not long afterwards we shot a rapid, the river having still further narrowed. The barking crow (*Corvus americanus*) is not seen to the northward of this place. In the *Fauna Boreali-Americana,* I have stated that it does not range beyond the 55th parallel; but more correct information, received on the present voyage, enables me to carry its northern limit on to the 61st. It

becomes rare before it ceases altogether to be
seen, and we have not noticed it in flocks since
leaving the Saskatchewan. In its gregarious habits
on the latter river it resembles the European rook,
but differs from that bird in the care with which
it conceals its nest. In the evening we landed
to cook supper, and afterwards re-embarked to drift
with the stream. At midnight, having come
to the Little Lake, where there is no current,
we could no longer drive; we therefore anchored
under a small sandy island, and at 4 A.M. on

July 21*st*, resumed our voyage. Four hours
afterwards we landed at the outlet of the lake to
cook breakfast. The morning was close and hazy,
with distant thunder; and at 10 A.M. the storm
approaching us, we were driven to take shelter for
a time under the bank of the river. When the
squall abated, we continued our voyage, notwithstanding that the rain fell throughout the day;
and during the night we again drifted with the
stream, the crews sleeping in the boats.

We made sail on the 22nd, at a quarter before
3 A.M., with a fair wind, which soon afterwards
chopped round against us, and increased to a
fresh breeze. At an early hour we passed the
mouth of Trout River; and after breakfast descended the westerly reach below the site of the
old fort. An hour later we passed the River La

Câche, and in an hour and a half more came to Hare-skin River. The rate at which we passed the land must have been at least seven geographical miles an hour; but the distances in this part of Sir John Franklin's chart are too great, and Fort Simpson, which was laid down by him from dead reckoning, is placed twenty miles too far north.

The river having, through the increase of the wind, become too rough for the use of oars, we worked down under sail, and made good progress, arriving at Fort Simpson at five in the afternoon. The position of this place, as ascertained by Mr. Thomas Simpson in 1836, is in latitude 61° 51′ 25″ N.; and longitude, deduced from lunar distances, 121° 51′ 15″ W.*

Between Desmarais's Fishery, on Slave Lake, and Fort Simpson, the direct distance is about one hundred and fifty-five geographical miles. In the

---

* From this it appears that, by some means, an error of twenty miles of latitude had crept into the reckoning of Sir George Back and Lieut. Kendall in 1825, between the old fort, in long. 120°, where the latitude was obtained by these officers, and Fort Simpson; but, on the other hand, they assigned too little departure, so that the mistake was in the courses as much as in the distance. And in correcting the chart, to give Fort Simpson its proper geographical position, a corresponding alteration must be made in the course and length of the river between that fort and the great bend below it, where the latitude and longitude were again ascertained by the observations of Back and Kendall.

wider parts of the river the coast is shelving, and not easily approached, in boats, from the shallowness of the water; but in the narrower places the beach is steep, and the channel is full of boulders. In the few spots where sections of the strata are visible, a bituminous shale, containing many fragments of the small pteropodous shell *Tentaculites fissurella*, indicates the formation to be the same with that on the Athabasca River and Slave Lake, which has been said above to be probably the Marcellus shale. Between the old fort and Hare-skin River, the basis of the bank is formed of a greyish green slate-clay, which, under the influence of the weather, breaks into scales like wacké, and at last forms a tenacious clay. The whole banks of the river seem to belong to a shale formation; but from the want of induration of the beds, they have crumbled into a slope more or less steep, and the capping of sand, clay, and boulders has fallen down and covered the declivity. On the south, a long even rising ground, named the Trout Mountain, which runs parallel to the river at a distance of from ten to twenty miles, is visible at intervals the whole way; and a similar but higher range, named the Horn Mountain, exists on the north.

Of the composition of these eminences, I have no information; but I suspect, from the evenness

of their outlines and their relative position, that
they are escarpments of the sandstone and shale
of the Erie group, remaining after the excavation
of the valley of the river, such as has been already
noticed as existing in the Clear-water and Elk
Rivers, and as we shall afterwards have occasion
to mention, when describing the north-west side of
Great Bear Lake.

The bank of the river at Fort Simpson is precipitous, and about thirty feet high ; but the river
sometimes flows over it in the spring floods, occasioned by accumulations of drift ice.　It is composed of sand and loam, and the beach is lined
with boulders of granite, greenstone, limestone, and
sandstone.

Barley is usually sown here from the 20th to
the 25th of May, and is expected to be ripe on the
20th of August, after an interval of ninety-two days.
In some seasons it has ripened on the 15th.　Oats,
which take longer time, do not thrive quite so well,
and wheat does not come to maturity.　Potatoes
yield well, and no disease has as yet affected them,
though the early frosts sometimes hurt the crop.
Barley in favourable seasons gives a good return at
Fort Norman, which is further down the river;
and potatoes and various garden vegetables are also
raised there.　The 65th parallel of latitude may,
therefore, be considered as the northern limit of

the *Cerealia* in this meridian ; for though in good seasons, and in warm sheltered spots, a little barley might possibly be reared at Fort Good Hope, the attempts hitherto made there have failed. In Siberia it is said that none of the corn tribe are found north of 60°. But in Norway barley is reported to be cultivated, in certain districts, under the 70th parallel. It takes three months usually to ripen on the Mackenzie, and on our arrival at Fort Simpson we found it in full ear, having been sown seventy-five days previously. In October 1836, a pit sunk by Mr. M'Pherson, in a heavy mixture of sand and clay, to the depth of 16 feet 10 inches, revealed 10 feet 7 inches of thawed soil on the surface, and 6 feet 3 inches of a permanently frozen layer, beneath which the ground was not frozen.

A number of milch cows are kept at Fort Simpson, and one or two fat oxen are killed annually. Hay for the winter provender of the stock is made about one hundred miles up the river, where there are good meadows or marshes, and whence it is rafted down in boats. We met the haymakers, being three men, some hours before we reached the fort, on their way to cut the grass, which is a bent that grows in water. The hay will be brought down in September.

The fort stands on an island at the junction of

the River of the Mountains (*Rivière aux Liards*) with the Mackenzie. This large tributary originates in the recesses of the Rocky Mountains, by many small streams which, uniting, form two branches. Both branches rise to the westward of the higher peaks, and afford another of the many instances of streams of magnitude crossing the chain. By Dease's River, which is the westernmost affluent of the north branch, boats pass through the mountains, and gain, after much trying and perilous navigation, and some portages, the Pelly and Lewis, at the junction of which the Company have a post named Pelly Banks. Native traders travel thither twice in the season from Lynn Canal, situated to the north of the island of Sitka, on the 59th parallel. This inlet is frequented by the Hudson's Bay Company's steamers, and, in this present summer of 1848, Mr. Todd, captain of one of these steamers, forwarded letters and newspapers to Mr. Campbell, the officer in charge at Pelly Banks. One of the newspapers, published at Honolulu, which was sent on to Fort Simpson, was transmitted by Mr. M'Pherson to Fort Confidence in the winter, and gave us the first intelligence of the origin of the gold hunt in California, and of the migration within a few days of two thousand men from Oregon, and of most of the Company's servants at Fort Vancouver, on that exciting pur-

M 4

suit. Such unexpected channels does commerce open for the conveyance of intelligence, and had previous arrangements been made, we might, by the route across the Andes at Panama, the Atlantic steamers to California or the Sandwich Islands, and this northern way back again across the Rocky Mountain ridge, have had much more recent intelligence of our friends in Europe than we were destined to receive during our long winter residence on Great Bear Lake.

The Lewis flows from a large sheet of water, lying within the English boundary, but named the Russian Lake, because Mr. Roderick Campbell, who was the first officer of the Hudson's Bay Company who visited it, met there a party of Russian traders. The influence of these rivals in trade is supposed to have caused the attack made by the natives on Mr. Campbell's post in the winter of 1839, which resulted in the loss of three of his party by famine, and the narrow escape of the remainder from the same fate, as related in the narrative of Dease and Simpson's voyage (p. 173.). Mr. Campbell, undaunted by this calamity, renewed his journeys in the same direction, and, in consequence of an agreement that had then been made between the Hudson's Bay and Russian Fur Companies, with less hazard. His first post, named after himself, was on the Pelly, and at the supposed distance from Fort Halkett on the River of the Mountains of three hundred miles, by the winter route, which is usually as direct as the nature of the country will admit. From Campbell's post to the Forks or junction of the Lewis and Pelly, where the present fort is situated, the distance is reckoned at two hundred and forty miles on a south-west course. To retrace this length of way, the crew of a light canoe are said to consume twelve days on the tracking line, being at the rate of twenty miles a day, which is generally considered as but an indifferent day's work against the current. It is probable, however, that the river is very tortuous, and that there are many impediments in a stream flowing through so mountainous a region. Of these two branches, the Lewis is the westernmost, and the river formed by their junction, which retains the name of Pelly, falls into the Pacific. By observations made by Mr. Campbell on the temperature of boiling water at Pelly Banks, the height of that post above the sea has been esti-mated at 1314 feet.

After the union of its two arms, the River of the Mountains flows for a considerable breadth of lon·gitude on the 59th parallel, and near the middle of this part, at the influx of Smith's River, Fort Halkett stands. Fort Liard is situated lower down, after the river has made a sharp turn to the north, in its course towards the Mackenzie, which it joins at Fort Simpson. Though this post is more elevated than Fort Simpson, by at least one hundred and fifty feet, and is only two degrees of latitude to the southward, its climate is said to be very superior, and its vegetable productions of better growth and quality. Barley and oats yield good crops, and in favourable seasons wheat ripens well. This place, then, or the 60th parallel, may be considered as the northern limit of the economical culture of wheat.

It has been already mentioned that the *Wapiti* or *Wawaskeeshoo* of the Crees, the representative of the European red deer, does not range to the north of the River of the Mountains, and the same stream marks the northern limit of the American magpie, Say's grouse, and the white crane (*Grus americana*).

Mr. M'Pherson had most kindly set aside for me a cask of excellent corned beef, cured at the fort, and some bags of very fine potatoes raised at Fort Liard, with several other things which he knew would be serviceable at our winter residence. I left them in store, for Mr. Bell to embark when he came up, together with such supplies of iron-work and dried meat as the depôt could furnish, and to convey them to our future winter residence on Great Bear Lake. The boats were hauled up, their bottoms payed over with boiling mineral

pitch, and such other repairs made as were necessary. I had intended to give them additional
false keels at this place, to render them safer and
more weatherly at sea, and, with this view, had
long bolts and screws prepared at Portsmouth dock-
yard, to fit plates sunk in the keels; but the bolts
were unluckily left behind at Cumberland House,
Mr. Bell not being aware of the purpose for which
they were designed, and we could not spare time to
make others. All our preparations having been
made on the 23rd, we left the fort on the 24th at
5 A. M., and three hours afterwards had the first
sight of the Rocky Mountains. In nine hours we
were exactly opposite the end of the first range,
where the Mackenzie, seemingly to avoid the barrier
formed by the mountains, makes a sudden flexure
from a north-west course to a north-north-east
one.

Here I must interrupt the narrative for a little, to
give some account of the geological structure of
the country through which the Mackenzie flows.

When the mountains are first seen in descending
the river, they present an assemblage of conical
peaks, rising apparently about two thousand feet
above the valley; and it is not until we come opposite to the end of the first mountain, that we
observe them to be disposed in parallel ridges
having a direction of about south-south-west and

Rocky Mountains at the bend of the River.

north-north-east *; which makes an angle of rather
more than forty-five degrees with the axis of the
great chain, from which they project like spurs.
The circumstance of the valleys pervading the chain
transversely, though with more or less of ascent,
explains the reason of the principal rivers on both
the eastern and western slopes having their sources
beyond the axis of the range, and flowing through it.
From some passages in Dr. Hooker's letters, I infer
that the Himalayas have a similar configuration.

As the successive spurs and the valleys between
them open out to the voyager who descends the

---

* I have never had leisure to ascertain the true course of
these ranges within six or seven degrees, but from the bearings
I have taken several times in passing I suppose that south 20°
west, and north 20° east is very near their direction.

river, he observes that the eastern faces of the ridges rise abruptly like a wall, while their western flanks are more shelving. This is not, however, uniformly the case, as in some of the ridges lofty escarpments occur also on their western sides.

The height of the almost precipitous cliff of the first mountain at the bend of the river appeared to the eye, from a distance of seven or eight miles, to be eight or nine hundred feet, though the width of the base of the hill did not exceed a mile. Further back, the summit of the ridge terminated by this mountain was judged to be between two thousand and two thousand eight hundred feet high. The heights here mentioned were estimated solely by the eye, and as in this climate heights and distances are very deceptive they must be considered as very rough approximations. No trees could be detected on the summits when examined with the telescope, but the lower hills, and the slopes to the height of a thousand feet, were well wooded.

The first range re-appears on the east side of the river, and is seen at intervals running in the direction of M'Vicar's Bay of Great Bear Lake, whose basin interposes between its termination and the granite and gneiss that skirt the eastern arms of that lake.

At the bend of the Mackenzie, the valley which

interposes between the first and second ridges does not appear to exceed five miles in width, but it was seen too obliquely to enable us to form a correct judgment. The river flows through this valley for upwards of fifty miles, when, making a small bend to the westward, it escapes across the ridge. Thus far the second ridge * runs on the west bank of the river, showing a bold precipitous craggy side at intervals, some parts being concealed from the voyager by the intervening swelling grounds which form the floor of the valley. Where the river cuts it, a high island of limestone stands in mid channel, and on the east bank, a round-topped hill, named the " Rock by the River's Side " † (*Roche qui trempe à l'eau*), rises precipitously from the water's edge to the height of five or six hundred feet or more. The base of this hill scarcely exceeds a mile in diameter, and most of the ridges seem to be of similar breadth. From the Rock by the River's Side the ridge continues, but with interruptions, onwards in the same direction to the elevated promontory of Great Bear Lake, named *Sas-choh etha* (Great Bear Hill), which stands between Keith's and M'Vicar's Bays.

The other spurs, which succeed these down to the delta of the river, rise in like manner like rugged walls from the surrounding low, undulating country,

---

* Partly seen on the right-hand side of the woodcut, p. 172.
† See woodcut, p. 182.

the stream escaping through them by successive gaps. Many of the escarpments, when seen from a distance reflecting the rays of the sun, look as bright and white as chalk cliffs; and but for information which I have gleaned from voyagers who have crossed them, I should have been in doubt whether they were not formed of that material or of white sand, instead of being hard limestone.

At this date only a few patches of snow remained in the hollows having a northern exposure; but in the following year they were entirely covered with snow until late in June, and for some weeks after all the low country had become quite bare. Both the first and second ridges are distinctly stratified at the bend of the river, and seemingly capped with trap. Where they and the succeeding ridges are cut by the river, limestone is the chief rock that is visible; but I have had no opportunity of examining the principal cliffs, and have made but a very cursory inspection of any. The spurs which reach the Mackenzie consist, perhaps, wholly of limestone. Sandstone exists in their vicinity, but I believe it is a newer deposit, belonging to that which forms the floors of the valleys, and rests unconformably on the tilted beds of the ridges. No organic remains were detected in any of the highly inclined beds, but gypsum and chert are of frequent occurrence.

Traders who have crossed from the Atlantic to the Pacific slopes of the continent say that there are fourteen or fifteen ranges of hills, and that when they are viewed from the summit of a peak, the mountain tops appear to be crowded together in great confusion, like a sea of conical billows. My informants could not tell me whether granite, clay-slate, or trap rocks entered into their composition or not; but it is probable that such is the case, as we know it to be in more southern latitudes. I received specimens of semi-opal, plumbago, and specular iron, gathered on one of the ridges. The more westerly ranges have obtained from the traders the name of the Peak Mountains.

On the Mackenzie, a shaly formation makes the chief part of the banks, and also much of the undulating valleys between the elevated spurs. It is based on horizontal beds of limestone, and in some places of sandstone, which abut against the inclined strata of the lofty wall-like ridges, or rest partially on their edges. Covering the shaly beds, there exists in many places a deposit of sand, sometimes cohering so as to form a friable sandstone; and where a good section of the bank occurs, a capping of gravel and boulders, of various thickness, is seen crowning the whole. The shale crumbles readily, and often takes fire spontaneously, occasioning the ruin of the bank, so that it is only by the encroachments of the river carrying away the debris that the true structure is revealed. The boulders that have dropped from above pave the beach in many places as closely and regularly as if it were a work of art, the passage of ice over them driving them firmly and evenly into the bed of tenacious clay which the shale in breaking down produces.

I have no evidence whereby the geological age of the shale may be certainly deduced, but am inclined to consider it as belonging to the epoch of the Marcellus deposit, on account of its exact lithological resemblance to the bituminous beds of Athabasca River, and the occurrence of the *Tentaculites fissurella* in the fragments which line the beach at the west end of Great Slave Lake.* The difficulty of deciding upon the age of the beds through which the river flows is increased by the occurrence among them of a tertiary lignite formation, which also takes fire spontaneously. This general account of the rocks of the Mackenzie is here introduced to facilitate the subsequent descriptions of such points as I landed upon.

With respect to some of the more remarkable quadrupeds that inhabit the Rocky Mountains, I

* In 1826, Mr. Sowerby referred some fossils which I obtained from the limestone beds of the Mackenzie, to the Oxford oolíte and cornbrash. These, which were mostly terebratulites, are not now within my reach, but should his opinion be confirmed by further specimens from the same quarter, they would indicate that the bituminous shale of the Mackenzie belongs to the lias.

may state that the mountain sheep, or big-horn as it is named (*Ovis montana*), frequents the higher peaks down to the delta of the Mackenzie. The Slave Indian appellation of this fine animal is *Sass-sei-yeuneh*, or "Foolish Bear." It keeps to the craggy summits, and can scarcely be approached by the hunter who ascends towards it from below; but should he once get above it, he can come near it easily. Its flesh is said to be equal to well fla-voured mutton, but its coat resembles that of the rein-deer, and is not woolly. The goat-antelope (*Antilocapra americana*), which is covered with a fine long-stapled wool, has its northern limit on the River of the Mountains. Its flesh is much inferior to that of the mountain sheep. Rein-deer, of a much larger size and darker colour than the "Barren-ground variety," frequent the mountain valleys; and moose deer, extending their range nearly to the Arctic Sea, through the wooded districts only, feed on the banks of the rivers where willows grow. Neither musk-oxen nor bison inhabit this part of the Rocky Mountains; the latter, as has been mentioned, having their northern limit on the Horn Mountain; while the former keep within the arctic circle, and to the east of the Mackenzie. - The little *Pika*, or tail-less hare, occupies the grassy emi-nences, and lays up a stock of hay for winter use. Say's grouse (*Tetrao Sayi*), named *Ti-choh*, i. e. "big grouse," has not been killed further north than the Nòhhané Bute ; the pin-tailed grouse goes as far down as the delta ; and the *Tetrao canadensis* lives in the marshy parts of the forest up to Peel's River, and is named *Ti ;* while the willow and white-tailed ptarmigans bear the designation of *Kasbah* or *Kampbah,* in the Slave or Chepewyan tongue. The last named is exclusively an Alpine species. The American magpie has not been seen to the north of the River of the Mountains, and is rare even there.

Many large streams join the Mackenzie below Fort Simpson. One, which the Nòhhanè Indians are accustomed to descend, flows down the valley between the first and second mountain ridges, and joins the Mackenzie at its great bend. It is designated from these people, but it must not be confounded with the stream of the same name, which issues also from the hunting-grounds of the Nòhhanès, but falls into the River of the Mountains.

The Willow Lake River enters the Mackenzie a little below the bend, from the right bank. It is ascended by the Marten Lake Indians as far as it is navigable for their canoes, and then a march of four hours, or of from ten to fifteen miles, takes them to Marten Lake.

Another river of considerable size comes in on the left bank, which is named the *Bekka-tess* by the *Dahadinnès* who frequent its banks, and *La Rivière de Gravoir* by the voyagers. It joins the Mackenzie in latitude 64½° N., and is said to issue from a large lake, situated on the summit, or even on the western side, of the Rocky Mountain range. The impediments to its navigation have prevented it from being used as a channel for the Company's trade; and it has been as yet only partially explored, though it has been thought that a route might be discovered through it to the banks of the Yukon. The *Dahadinnès* speak a dialect of the Chepewyan tongue. Mr. M'Kenzie, the gentleman in charge of Fort Norman, to which these people resort, informed me that their correct designation in their own language is *Cheta-ut-tinné* or *'Dtcheta-ta-ut-tinne*, which, being also the national name of the Beaver and Strong-bow or Mountain Indians, points them out as members of the same nation.

CHAP. VI.

WE drifted with the stream all night, and in the
morning of July the 25th, a thick fog preventing
us from pulling, we continued to drift, trusting
that the current would carry us clear of shoals and
low islands.   The sky cleared at breakfast-time,
and by noon we were abreast of the " Rock by the
River's Side."   In some places, where there are
islands, the river is two or three miles wide; in
others, it does not appear to be more than a mile,
or a mile and a half.   The small island, which lies
in the channel just above the Rock by the River's
Side, is composed of blackish-grey compact limestone, dipping to the south-half-east, at an angle of
about twenty degrees; the upper bed, which is
thinner and more slaty than the others, being
composed of irregularly oblong distinct concretions.
On the upper or south side of the Rock by the
River's Side the stone is a bituminous limestone,
yielding the smell of *stinkstein* when struck: the

N 3

precipitous face of the rock appears to be the same
kind of limestone.   Immediately below the Rock,
for the distance of half a mile, limestone similar to
that of the small island occurs in gently inclined
beds.

Rock by the River's Side.

In the body of this high bluff the beds are
nearly vertical; and, as well as I could judge from
the view obtained in descending the stream, they
were disposed as if the axis of the ridge had been
the direction of the elevating force, the beds inclining towards the summit from both sides.   In
some parts, there seemed to be inclined beds lying
nonconformably over the ends of the nearly vertical ones, but I could not be certain, without
closer examination, that what I saw was not merely
oblique sections of the edges of the lower beds.

A thermal spring, much resembling sea-water in its saline contents, issues from the front of the cliff, and the fissure from whence it flows is incrusted with crystallised gypsum.* Shale beds abut against the lower side of the Rock, covering the limestone beds above mentioned; but they are in a great measure concealed by the shelving debris of the bank. Contiguous to the upper or south side of the Rock there are sloping banks of gravel, capped by a vertical wall of friable sandstone. And three miles higher up the stream, there are two river terraces, more complete than any I noticed elsewhere on the Mackenzie, though in many places a high and low bank can be traced. These terraces are composed of fine sand; and the slope between them is so steep as to require to be ascended on all fours. Both terraces are very regular in their outlines, and are covered with well grown *Pinus banksiana.* The uppermost is about two hundred and fifty feet above the river. From this terrace, the Rock by the River's Side is clearly seen to be part of a chain, which is crossed there by the river, as has been already mentioned. This is not so evident from the channel of the stream. The high sand-banks continue almost with-

* Dr. Davy, who kindly analysed some water from this spring, ascertained that the chief saline ingredient was sulphate of magnesia.

N 4

out a break for twenty miles further up, and in some places they are seen to rest upon a grey shale. At one place, where there is a good section, it was perceived that the surface of the shale on which the sand reposed was uneven, and much indented also by pot-holes and projecting tongues; the gravel and sand descending into the pits, and the points of shale rising among the sand. The similarity of these shale and sand-cliffs to those at the junction of the Clear-water and Elk Rivers is very great; but the shale generally is not so bituminous as at the latter locality. The surface of the country above is strewed with gravel and boulders, and in the decay of the bank these fall down and line the channel of the river. When the water is high, as it is in the spring, little flat beach is to be seen; but in the autumn, the pavement of boulders to which I have already alluded is exposed. Among these, above the Rock by the River's Side, I observed a considerable number of granites, some gneiss, many sienites, basalts, and greenstones; also felspar rock, felspar porphyries, Lydian-stones, quartz rock, and limestones of various kinds, with quartzose sandstones, white, red, and spotted.

I have been disposed to give a more full abstract of the notes I made in descending and ascending this part of the river, because, in following its oblique course of more than fifty miles, from the first ridge of the Rocky Mountains at the bend, to the second at the Rock by the River's Side, all the various strata of the valley are seen, and, if properly examined, there is little reason to doubt that a key to the geological formations of the entire length of the Mackenzie might be obtained.

On the left bank, six miles below the Rock by the River's Side, beds of shale appear, having a slight dip to the southward ; and the ridge, which is prolonged on that side from the rock above-named in a north-north-east direction, appears very rugged, with irregularly serrated summits, the crest being apparently extremely narrow. The country between the ridges seems to be pretty even, except where it is cut by rivulets ; and the high bank of the river is level, though in places it looks hummocky or hilly, because of the gullies which intersect it.

In the evening we landed to cook supper at the mouth of Black-water River, which issues from a lake of the same name lying on the eastern bank ; and, embarking again to drift during the night, passed a bend of ninety degrees, which the river makes to the westward, and which is known to the voyagers by the appellation of " The Angle " (*L'équerre*). It marks the passage of the river through another range, of which a high hill on the eastern bank, named Clark's Hill, is the most conspicuous part. The ridge continued from this hill crosses Bear Lake River in the middle of its course, and there forms a rapid.

A short way below the "Angle," the Red Rock River, named also *Rivière des Grosses Roches*, flows in from the west. It looks wide at its mouth, but is not a large stream. Fifteen miles further down, the Gravel or Dahadinnè River, already mentioned, flowing also from the mountains on the left, comes in below the site of an old fort. We were opposite to this when we resumed our oars on the morning of the 26th at four o'clock, and soon afterwards, passing a sandy promontory on the left hand named the "Crumbling Beaver" (*Castor qui déboule*), we arrived at Fort Norman. Obtaining here a bottle of milk as a grateful addition to our breakfast, we landed two hours later to prepare that meal, and at noon reached the mouth of Bear Lake River. Between Fort Norman and this river a tertiary coal formation occurs, which deserves particular notice.

The *coal*, when recently extracted from the beds, is massive, and most generally shows the woody structure distinctly, the beds appearing to be composed of pretty large trunks of trees lying horizontally, and having their woody fibres and layers much twisted and contorted, similar to the white spruce now growing in exposed situations

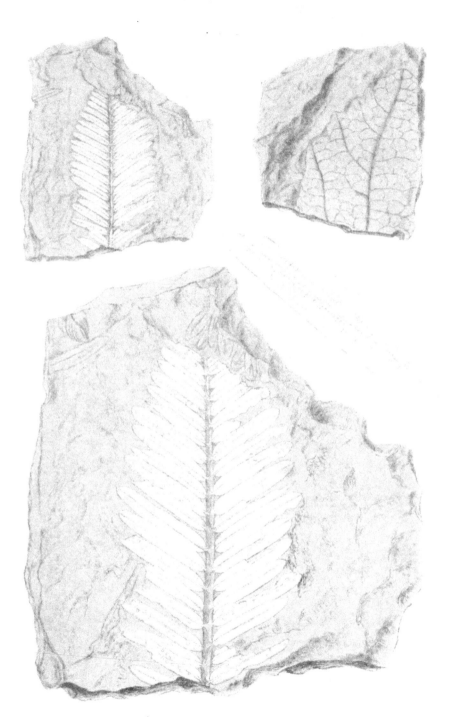

in the same latitude. Specimens of this coal examined by Mr. Bowerbank were pronounced by him to be decidedly of coniferous origin, and the structure of the wood to be more like that of *Pinus* than *Araucaria* ; but on this latter point he was not so certain. It is probable that the examination of a greater variety of specimens would detect several kinds of wood in the coal, as a bed of fossil leaves connected with the formation reveals the existence at the time of various dicotyledonous trees, probably *Acerineæ*, and of one which I am inclined to consider as belonging to the yew tribe. To these I shall refer again.

When exposed for even a short time to the atmosphere, the coal splits into rhomboidal fragments, which again separate into thin layers, so that it is difficult to preserve a piece large enough to show the woody structure in perfection. Much of it falls eventually into a coarse powder; and if exposed to the action of moist air in the mass it takes fire, and burns with a fetid smell, and little smoke or flame, leaving a brownish-red ash, not one tenth of the original bulk of coal taken from the purer beds, for some contain much more earthy matter.

Different beds, and even different parts of the same bed, when traced to the distance of a few hundred yards, present examples of " fibrous brown coal," "earth-coal," "conchoidal brown-coal," and "trapezoidal brown coal." Some beds have the external characters of "compact bitumen;" but they generally exhibit in the cross fracture concentric layers, although from their jet-like composition the nature of the woody fibres cannot be detected by the microscope. Some pieces have a strong resemblance to charcoal in structure, colour, and lustre. Very frequently the coal may be named a "bituminous slate," of which it has many of the lithological characters, but on examination with a lens it is seen to be composed of comminuted woody matter, mixed with clay and small imbedded fragments resembling charred wood. Crystals of selenite occur in this slate, and also minute portions of resin, or perhaps of amber. When this shaly coal is burnt, it leaves light, whitish-coloured ashes. The shape of the stems and branches of the trees is best preserved when they contain siliceous matter or iron-stone; and in this case, the bark of the tree is often highly bituminised, and falls off from the specimen.

From the readiness with which the coal takes fire spontaneously, the beds are destroyed as they become exposed to the atmosphere; and the bank is constantly crumbling down, so that it is only when the debris have been washed away by the river, that good sections are exposed. The beds were on fire near Bear River, when Sir Alexander Mackenzie discovered them, in 1785, and the smoke, with flame visible by night, has been present in some part or other of the formation ever since.

From one to four beds of coal are exposed above the water level on the banks of the river, the thickest of which exceeds three yards, and was visible a short way above Bear River in the autumn only,—the Mackenzie being then seven or eight feet below its spring level.

Interstratified with the coal beds, there are layers of *gravel* which occasionally, through the intermixture of clay more or less iron-shot, acquire tenacity enough to form vertical cliffs, but more often are very crumbly. The pebbles composing the gravel vary in size from that of a pea to that of an orange, and are formed of Lydian-stone, flinty slate, white quartz, quartzose sandstone and conglomerate, clay-stone, and slate-clay. The gravel is sometimes seamed by thin layers of fine sand, and its beds vary in thickness up to thirty or forty feet.

In place of the gravel, a friable *sandstone* is often interposed between the coal beds or rests upon them. It is fine-grained, often dark from the dissemination of bituminous matter, and has so little tenacity, that in many places it is excavated by the sand-martens. Being porous, it fills with water, and is frozen into a compact, hard rock, for most of the year; but becomes moist, and breaks down under the influence of the hot rays of the sun in spring.

Potter's clay, of a grey or brown colour, alternates with the beds already named, in layers varying from one foot to forty or more in thickness. This clay is often highly bituminous, and is penetrated by ramifications of carbonaceous matter, resembling the roots of vegetables. About ten miles above Great Bear River, a layer of this material, lying immediately over a bed of coal which was on fire, has been baked so as to resemble a fine yellowish-coloured *biscuit porcelain*. In a part of this, I found numerous impressions of leaves, most of them dicotyledonous, but one of them apparently coniferous, and belonging, probably, to the yew genus. The existence on many of the leaves of the latter plant of little round bodies like the fructification of ferns, invested the specimens with much interest: and I am indebted to Mr. Brown for examining them, and superintending the accurate drawings, made by Mr. Sowerby, junior, from which the accompanying plates have been engraved. The clay had unfortunately cracked so much under the influence of the heat to which it had been subjected, that I could not obtain entire specimens of the larger dicotyledonous leaves,

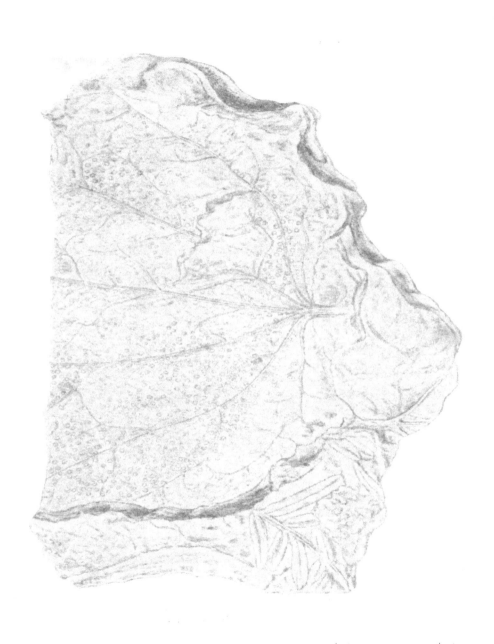

but in the general character of their venation they resemble the *Acerineœ*. Some portions of the clay was semi-vitrified, and so hard as to receive no impression from a file ; and I gathered pieces of this kind, composed of blue semi-vitrified layers, alternating with others of a rich buff colour. All the indurated clay, containing leaves, splits easily into thin layers, in every one of which there were impressions, so that the various kinds of leaves must have been deposited thickly above one another at this place. The fossiliferous clay is covered by one hundred and forty feet of sand and sandstone, and by some thin layers of conglomerate.

A *pipe-clay* is very generally associated with the coal beds, and is frequently found in contact with the lignite. It exists in beds varying in thickness from six inches to a foot, and is generally of a yellowish-white colour, but in some places has a light lake-red tint. It is smooth, without grittiness, and when masticated has a flavour somewhat like the kernel of a hazel nut. When newly dug from its bed, it is plastic, but in drying becomes rather meagre and adheres to the tongue : its streak is less glistening than that of ordinary English pipe-clay. As the natives eat this earth in times of scarcity, and suppose that thereby they prolong their lives, I requested Dr. Davy and the late Dr. Prout to examine it, but neither of these able chemists could detect any nutritious matter in it. Neither have I been able with the microscope to discover in it the remains of any infusorial animals.* Mr. Nuttall speaks of a similar substance under the name of *pink-clay*, which he observed in the lignite deposits on the Arkansa. It is known generally among the residents at the fur posts on the Mackenzie by the appellation of "white-mud," and is used for whitewashing houses, and also, when soap is scarce, for washing clothes.

In one place in the vicinity of the burnt cliff where the leaves were found, several beds of *porcelain-earth* occur from two to three yards thick, and apparently replacing the sandstone of other parts of the formation. It has a whitish colour, and at first sight looks like chalk, but some of its beds have a greyish hue from the quantity of carbonaceous matter disseminated in them. Its texture is fine-granular; it adheres slightly to the tongue, yields readily to the nail, is meagre, and soils the fingers slightly. Besides the coaly matter, it contains, also, a few minute scales of mica, and some of quartz. It is not plastic, and becomes more friable when moistened with water; neither does it effervesce with acids. This lignite formation extends

* Baron Humboldt mentions a tribe of Indians residing on the Orinoco, who eat large quantities of clay when food is scarce.

from the Rocky Mountain spur of which Clark's Hill forms a portion, to the spur of which the hill on the lower side of the mouth of Great Bear River is a prominent point, being directly across the valley about twenty-five miles, but considerably further by the course of the Mackenzie. The depth to which the deposit descends below the bed of the Mackenzie was not ascertained, but the height from the surface of the water to the top of the bank varies from ninety to one hundred and fifty feet. Ten or twelve feet or more of the crest of the bank consists of diluvial gravel with boulders, and the soil is generally peaty to the depth of a foot or two. The beds are usually almost horizontal or have a very moderate inclination, but in some few places they dip very considerably, and in the second reach of the Mackenzie above Great Bear River a bed of stone passes obliquely from the top to the bottom of the clay bank. By the destruction of the coal beds the cliff falls down, the slope is covered with the gravel and boulders, and the latter pave the channel of the river also. The strong current of the river varies its direction from time to time, and as the deposition or removal of alluvial islands expose or protect the banks, the debris of the ruined cliffs accumulates or is carried away. This constant waste of the bank would proceed much more rapidly, were it not that the ground is still frozen hard when barriers of ice, during the high spring floods, often raise the river thirty feet above its ordinary level. Then the frozen earth resists the action of the water as a rock would do, and the surface yields only in proportion as it thaws, which is slowly, since the water loaded with ice is kept down to the freezing point.

I observed that the bank of the river was generally higher than the land behind it, by at least the thickness of the diluvial capping, and sometimes by a part of the sand or clay of the tertiary beds, and that the narrow elevated bank extended in the same form along the principal affluents, a marked instance of which occurs on the south side of Great Bear River. In consequence of this configuration of the surface, the spring floods of melting snow accumulate, and at length make their escape through gullies, contributing further to the ruin of the bank, and giving it a broken and hilly outline when seen from the river. Landslips are of common occurrence, and are occasioned by pressure of water collecting in fissures produced by the partial subsidence of the cliff.*

---

* Similar tertiary coal formations occur on the flanks of the Rocky Mountains; the most southerly one of which I have any account, being in the Raton Pass, in latitude 37° 15′ N., longitude 104° 35′ W., and upwards of seven thousand feet above the level of the sea. Leaves of dicotyledonous trees, obtained in these beds by Lieutenant Abert in 1847, are figured

The Mackenzie traverses the basin in which the tertiary coal is deposited very obliquely, and the Great Bear Lake River cuts it more directly across.

in Colonel Emory's report to Congress (pp. 522. 547.). Nuttal observed lignite beds associated with the pink-coloured pipe-clay on the Arkansas, somewhere near the 48th parallel. Sir Alexander M'Kenzie states that a narrow strip of marshy, boggy, and uneven ground, producing coal and bitumen, runs along the eastern base of the Rocky Mountains, and he specifies latitude 52° N., longitude 112½° W., on the southern branch of the Saskatchewan, and latitude 56° N., longitude 116° W. (Edgecoal Creek) in the Peace River, as places where coal beds are exposed. Mr. Drummond procured me specimens of coal with its associated rocks at Edmonton (latitude 53° 45′ N., longitude 113° 20′ W.) on the north branch of the Saskatchewan, and, consequently, between the places mentioned by Sir Alexander M'Kenzie. According to Mr. Drummond the coal was in beds varying in thickness from six inches to two feet, and interstratified with clay and sandstone. The examples he selected were precisely similar to the slaty and conchoidal varieties which are found at the mouth of Great Bear River, and the resemblance between the sandstone of the two localities is equally close. He also found a black tertiary pitch coal which breaks into small conchoidal and cubical fragments, which Mr. Small, a clerk of the Hudson's Bay Company, who gave me the first information of these beds, likened well to Spanish-liquorice. At Edmonton the more slaty coal-beds pass gradually into a thin, slaty, friable sandstone, which is much impregnated with carbonaceous matter, and contains fragments of fibrous lignite. Hand specimens of this cannot be distinguished from others gathered from the shale cliffs on the Athabasca River. Highly bituminised shale, considerably indurated, exists in the vicinity of the coal at Edmonton, and clay-ironstones occur in the clay beds.

Chief Factor Alexander Stewart told me that beds of coal

The hill on the north side of the last-named river rises about six or seven hundred feet above the water, every where steeply, and in some places precipitously. It is, as has been stated, part of one of the spurs of the Rocky Mountains, a gap in which furnishes a channel for the passage of the Mackenzie. Its base, measured directly across, is about three quarters of a mile. The Great Bear River flows between its south flank and the tertiary coal beds described above; but on its north flank horizontal beds of limestone and bituminous shale appear again. The strata of the hill itself are highly inclined upwards on both its flanks towards its axis, and some are vertical. I did not procure organic remains from any of the upheaved beds forming these ridges or spurs, whereby their age might be determined, but they are evidently older than the limestone and shale formation which abuts against them or covers their edges, and are

are on fire on the Smoking River, which is a southern affluent of the Peace River, and crosses the 56th parallel of latitude, and also that others exist on the borders of Lesser Slave Lake, that lies between Smoking River and Edmonton. There are coal beds on fire, also, at the present time near Dunvegan on the main stream of the Peace River. All these places are near the base of the Rocky Mountains, or the spurs issuing from that chain, and their altitude above the sea varies from 1800 to 2000 feet and upwards. The beds at Great Bear River are probably not above 250 feet above the sea level.

very probably, judging from the scarcity of fossils, of the protozoic epoch.

Hill at the Rapid on Bear Lake River.

The Hill at the Rapid, twenty-four miles higher up Bear Lake River, is very similar to the one just noticed, and its beds have the same anticlinal arrangement. It is, as has been already stated, a member of the same spur with Clark's Hill, and from its summit the ridge may be seen extending through a comparatively level country towards the west end of Smith's Bay in Great Bear Lake. The floor of the valley lying between it and the spur at the mouth of the river is well wooded, but is much intersected by lakes, marshes, and considerable streams, some of which fall into the Mackenzie, and others into Bear Lake River. Immediately to

o 3

the westward of the Hill at the Rapid, but separated from it by a rivulet, there are horizontal beds of friable sandstone, and beyond them a thick deposit of bituminous shale, which extends northwards into the high promontory of the Scented Grass Hill, that divides Smith's Bay from Keith's Bay in Great Bear Lake.   The excavation of the body of the lake terminates the shale formation in this direction, but more to the westward it can be traced onwards to the Arctic Sea.*

* Various detailed accounts of some of the tertiary coal beds, and of the elevated spurs which cross Bear Lake River, are contained in the Geological Appendix to Franklin's Second Overland Journey; and the maps on a large scale, given in that work, may be consulted with advantage by any one who wishes to become well acquainted with the topography of the country, or to trace the course of the ridges here described in the text.

The limestone which forms the body of the hill at the mouth of Great Bear Lake River is blackish grey, full of sparry veins, or brownish-grey and bituminous, associated with calcareous breccia.   On the northern flank of the hill, abutting against the vertical beds, there are layers of bituminous shale, some of which effervesce with acids, while others approach in hardness to flinty slate.   Underlying the shale, horizontal beds of limestone are exposed for some miles along the Mackenzie, and from them there issue springs of saline sulphureous waters and mineral pitch.

The horizontal sandstone beds, above the Hill at the Rapid, of the same river contain fossils, some of which were considered by Mr. Sowerby to belong to the same age with the English oolitic limestones; but they require re-examination, and then we may learn whether the very extensive bituminous formation belongs to the Marcellus shale or to the lias beds.

As has been already said, the general aspect of the forest does not alter in the descent of the Mackenzie. The white spruce continues to be the chief tree. In this quarter it attains a girth of four or five feet, and a height of about sixty in a growth of from two to three hundred years, as shown by the annual layers of wood. One tree, cut down in a sheltered valley near Clark's Hill, measured the unusual length of one hundred and twenty-two feet, but was comparatively slender. Most of the timber is twisted, particularly where the trees grow in exposed situations. The Banksian pine was not traced to the north of Great Bear Lake River; but the black spruce, in a stunted form, is found on the borders of swamps as far as the woods extend. The dog-wood, silvery oleaster (*Elæagnus argentea*), *Shepherdia*, and *Amelanchier* grow on banks that in Europe would be covered with gorse and broom, and the southern *Salix candida* is replaced by the more luxuriant and much handsomer *Salix speciosa*, which is the prince of the willow family. The *Hedysarum Mackenzii* and *boreale* flower freely among the boulders that cover the clayey beaches ; while the showy yellow flowers and handsome foliage of the *Dryas Drummondii* cover the limestone debris, which give shelter also to the *Androsace Chamæjasmi*. In the heart of the spruce-fir forests, the curious and beautiful *Calypso borealis* lurks, along with some very fine, large, one-flowered, ladies' slippers (*Cypripedia*). There is, in fact, notwithstanding the near neighbourhood of the arctic circle, no want of flowering plants to engage the attention of a student of nature; and many of the feathered inhabitants of the district recall to the traveller or resident fur-trader pictures of southern domestic abodes. The cheerful and familiar *Sylvia æstiva* is one of the earliest arrivals in spring, coming in company with the well-known American robin (*Turdus migratorius*) and the purple and rusty grakles. A little later, the varied thrush makes its appearance from the shores of the Pacific. The white-bellied swallow (*Hirundo bicolor*) breeds, at Fort Norman, in holes of rotten trees; and the *Sialia arctica*, a representative of the blue bird so common in the United States, enlivens the banks of the Mackenzie, coming, however, not from the Atlantic coasts, but from the opposite side of the Rocky Mountain range. On the Mackenzie, there is an intermingling of the floras of both coasts, as well as of the migratory feathered tribes, the Rocky Mountain range not proving a barrier to either.

One of the birds which we traced up to its breeding-places on Bear Lake River, but not to the sea-coast, is the pretty little Bonapartean gull (*Xema Bonapartii*). This species arrives very early in the season, before the ground is denuded of snow, and seeks its food in the first pools of water which form on the borders of Great Bear Lake, and wherein it finds multitudes of minute crustacean animals and larvæ of insects. It flies in flocks, and builds its nests in a colony resembling a rookery, seven or eight on a tree; the nests being framed of sticks, laid flatly. Its voice and mode of flying are like those of a tern; and, like that bird, it rushes fiercely at the head of any one who intrudes on its haunts, screaming loudly. It has, moreover, the strange practice, considering the form of its feet, of perching on posts and trees; and it may be often seen standing gracefully on a summit of a small spruce fir.

The insectivorous habits of this bird, and its gentle, familiar manners, contrast strongly with the predaceous pursuits and voraciousness of the short-billed gull (*Larus brachyrhynchus* of the *Fauna Boreali-Americana*). If a goose was wounded by our sportsmen, these powerful gulls directly assailed it, and soon totally devoured it, with the exception of the larger bones. In the spring of 1849, when Mr. Bell and I were encamped at the head of Bear Lake River, waiting for the disruption of the ice, the gulls robbed us of many geese, leaving nothing but well-picked skeletons. Mr. Bell, who was the chief sportsman on this occasion, and spent the day in traversing the half-thawed marshes in quest of game, hung the birds, as he shot them, to the branch of a tree, or deposited them on a rock; but, on collecting the produce of his chase in the evening, he found that the gulls had left him little besides the bones to carry. If by chance a goose, when shot, fell into the river, a gull speedily took his stand on the carcase, and proceeded to tear out the entrails, and devour the flesh, as he floated with it down the current. Even the raven kept aloof, when a gull had taken possession of a bird.

The harlequin duck (*Clangula histrionica*) also frequents Bear Lake River; but is comparatively rare in other districts, and is not easy of approach. It congregates in small flocks, which, lighting at the head of a rapid, suffer themselves to glide down with the stream, fishing in the eddies as they go. A sportsman, by secreting himself among the bushes on the strand, conveniently near to an eddy, may, if he has patience to wait, be sure of obtaining a shot. In this way I procured specimens. The osprey and white-headed eagle both build their nests on the banks of Bear Lake River, and the golden-winged woodpecker migrates thus far north, and perhaps further, though it did not come under our observation in a higher latitude.

A small frog (*Bufo americanus*) is common in every pond, and Mr. Bell informed me that he had seen it on Peel River, which is the most northern locality I can name for any American reptile.* A frog resembling it, but perhaps of a different species, abounds on the Saskatchewan, and its cry of love in early spring so much resembles the quack of a duck, that while yet a novice in the sounds of the country, it led me more than once to beat round a small lake in quest of ducks that I thought were marvellously well concealed among the grass.

On Bear Lake River, the frogs make the marshes vocal about the beginning of June. Throughout Rupert's Land, they come abroad immediately after the snow has melted. In the swampy district between Lake Superior and Rainy Lake, they are particularly noisy. While we were descending the Savannah River on the 20th of May, we were exposed to the incessant noise of one called by the voyagers *le crapaud*†, whose cry has an evident affi-

---

* See note p. 204.

† This is probably the *Bufo americanus*, also. Mr. Gray of the British Museum, who examined my specimens, found old and young examples of *B. americanus* from Lake Winipeg, and young ones from Great Bear Lake. There were also many specimens of *Rana sylvatica* (*le grenouille*) from the former locality; some of *Hyla versicolor* of Le Conte, or *H. verrucosa* of Daudin; and a solitary individual of a *Hylodes*, which he thinks may be new. It resembles, he says, "*H. maculatus* of Agassiz (Lake Superior, p. 378. pl. VI. f. 1–3.), but differs in colour. The back is grey, with three cylindrical dark bands, interrupted and diverging from each other on the hind part of the back. The side of the face has a black streak, which is continued over the base of the fore-arm, and along the side of the body, gradually descending towards the belly. The toes are

nity with the *brekekex* of Asia Minor, and closely resembles the braying sound of a watchman's rattle; but a hundred of the latter, sprung in a circle, would not have equalled the voices of the frogs that we heard at one time. A smaller species, called *la grenouille*, inhabits the same places, and has a shrill, less unpleasing note than the other, yet which was, nevertheless, tiresome from its monotony.

As a contribution to what is known of the geographical distribution of reptiles, on the east side of the Rocky Mountains, frogs may be set down as attaining the 68th parallel of latitude; snakes, as reaching the 56th; and tortoises, as disappearing beyond the 51st, at the south end of Lake Winipeg. There the *Emys geographica* of Le Sueur, named *asatè* by the Chippeways, occurs; and also one with a flexible neck, called by the same people *miskinnah*, which is probably the snapping turtle.*

free and cylindrical, that is, scarcely tapering, and truncate at the end." (*I. E. Gray* in let.)

* By the same post which brought me a proof of this sheet, I had a letter from Mr. Murray, dated on the River Yukon, in which he informs me that "a frog" and "a grass snake" had been killed near his encampment, and that another snake had been killed on the north bend of the Porcupine River, far within the arctic circle.

CHAP. VII.

WE continued to descend the river until 7 in the evening, when we encamped for the night, as I did not consider it to be safe to drift here, there not being one person in the boats who had ever been in this river before, but myself, and I could not trust to my recollections of the best channels after the lapse of so many years since my former visit.

About twelve or fourteen miles below the influx of Great Bear River, the channel of the Mackenzie approaches the spur on its eastern bank, and flows parallel to it for some distance. At the spot where we encamped the beach was formed of dis- placed bituminous shale with imbedded granite boulders, both evidently derived from the ruined bank, a section of which showed layers of gravel consisting of rolled pieces of shale and a few limestone pebbles, alternating with sand and coarser rolled pieces of limestone. This seemed to be a tertiary deposit formed out of the subjacent beds, but not by the river flowing at its present level.

In the course of the day's voyage we noticed a peregrine falcon's nest, placed on the cliff of a sandstone rock. This falcon is not rare throughout the Mackenzie, where it preys on the passenger pigeons and smaller birds. Mr. M'Pherson related to me one of its feats, which he witnessed some years previously as he was ascending the river. A white owl (*Stryx nyctea*), in flying over a cliff, seized and carried off an unfledged peregrine in its claws, and, crossing to the opposite beach, lighted to devour it. The parent bird followed, screaming loudly, and, stooping with extreme rapidity, killed the owl by a single blow, after which it flew quickly back to its nest. On coming to the spot, Mr. M'Pherson picked up the owl, but, though he examined it narrowly, he could not detect in what part the death-blow had been received; nor could he, from the distance, perceive whether the peregrine struck it with wing or claws.

July 27th.——Embarking at 3 this morning, we continued our voyage down the river, and for upwards of twenty miles pursued a course nearly parallel to the spur which the Mackenzie crosses at the influx of Great Bear River. In latitude

65° 32′ N., longitude 127 W., we were opposite to a
magnificent cliff in this ridge, only two or three
miles inland, apparently about four hundred feet
high, and some miles in length.   The escarpment
faces directly southwards, is remarkably white, and
the layers composing it are nearly horizontal, but
with some undulation.   The heights of the peaks
appeared to me to be about eight hundred feet
above the water.   The beach is composed of fragments of bituminous shale with pieces of lignite;
and five or six miles further down, there is a good
section of the shale beds interstratified with dark-
coloured sandstone.

At the "Rapid" the Mackenzie crosses another
spur, making three elbows in its passage through it.
The channel of the river there is formed of limestone,
and is shallow, producing, when the water is low,
a considerable fall on the east side, and a shelving
rapid on the west.   At the elbow of the river, above
the rapid, one of the hills, which rises steeply from
the water's edge on the east bank, is composed of
limestone beds, wrapping over one another like the
coats of an onion, and curving, at the place where
this structure was most distinctly seen, at a spherical
angle of 65°, or thereabouts.   These inclined beds
are capped and covered on the flanks by strata
of sandstone, which breaks down readily and forms
a steep talus of pale-red sand.   A cliff of the upper

and more compact sandstone overhangs the crumbling layers beneath it.

Another eminence of the same spur, which rises from the rapid a few miles lower down, shows the same conical elevation with curved concentric beds. In one spot there is a fault, with dislocation of the beds. On both flanks of these inclined beds there are layers of aluminous shale interstratified with limestone and sandstone. Where these shale beds rest on the inclined rocks, they are also inclined, but they rapidly assume the horizontal position as they recede from the hill.

In the earlier part of the summer, a steam-boat could ascend the rapid without difficulty; and this great river might be navigated by vessels of considerable burden, from the Portage of the Drowned in Slave River, down to its junction with the sea, being a navigation of from twelve to thirteen hundred miles.

In a dilatation of the river, about ten miles below the rapid, bituminous shale lies horizontally in the hollows of undulated beds of limestone. Having cooked supper at this spot, we embarked to drift for the remainder of the night.

At 5 in the morning of the 28th, we were at the commencement of the Ramparts, where the river is hemmed in to the width of from four hundred to eight hundred yards, and has a strong current. This is the " second rapid " of Mackenzie, who states that it is fifty fathoms deep; but in obtaining such soundings, his lead must have fallen into a crevice, or have been carried down the channel of the stream by the strength of the current; for gentlemen of the Hudson's Bay Company, who are well acquainted with the locality, informed me that a bed of stone crosses the stream, and at the close of the summer, when the river is at the lowest, produces a fall, except on the east side, where there is a channel that boats can ascend by towing. In the dilatation of the river above the Rampart defile, there are some fine examples of sandstone cliffs, which have decayed so as to form caves, pillars, embrasures, and other architectural forms. The beds have slate-clay partings and seams of clay-ironstone. Associated with them there is a marly stone, containing corallines, referred by Mr. Sowerby to *Amplexus*; and covering the sandstone in many places, and alternating with the upper beds, there is a deposit of bituminous shale.

In making its way through the defile, the river bends suddenly to the east-north-east, and, as the dip of the beds forming the cliffs on each side is in the contrary direction, the strata rise into sight in succession as we descend the river. The cliffs have been denuded of the covering of shale which exists higher up the stream, but the limestone of which they are chiefly formed is stained with bitumen, either in patches or whole layers. At the upper end of the defile, a fine granular, foliated limestone is interleaved with beds containing madrepores, and parted by seams of carbonaceous matter. Near the middle of the defile the limestone contains the *Terebratula sphæroidalis* (or a nearly allied species) which is a fossil of the inferior oolite, also some *Producti* and the coralline named *Amplexus*. Several seams of black shale, about eighteen or twenty inches

Section at the Ramparts.

thick, exist among the limestones, and rise with them in succession above the level of the water ; but there are not more than two or three of these seams in the height of the cliff at any one part. The shale is of various degrees of hardness, and passes into a brownish-black flinty slate. The dip of the beds is not uniform throughout the defile, being more or less undulated, and for some way the layers are horizontal. In places here and there, the limestone beds are excavated into deep pot-holes filled with shale, resembling the gravel pits which dip into chalk beds.

On the top of the Rampart Cliffs we found a large body of Hare Indians encamped. This is a common summer haunt of these people, who resort thither to avail themselves of the productive fishery which exists above the defile. At this time, owing to the river not having subsided so rapidly as usual, they were taking only a small number of fish, and, consequently, were complaining of want of food. This people, and most of the tribes who live the whole year on the immediate banks of the Mackenzie, depend greatly for subsistence on the hare (*Lepus americanus*). Of these animals they kill incredible numbers; but every six or seven years, from some cause, the hares disappear suddenly throughout the whole country; so that not one can be found either dead or alive. In the following year a few reappear, and in three years they are as numerous as before. The Canadian lynx migrates when the hares, on which it chiefly preys, become scarce. The musk rat is subject to periodical murrains, when great numbers lie dead in their nests; but the dead hares are not found, whence we may conjecture that when their numbers become excessive they disappear by migration. I could not learn, however, that the Indians had ever seen them travelling in large bands.

The Hare Indians are a tribe of the Tinnè or Chepewyan nation, and speak a language differing

only as a provincial dialect. They are, like the
rest of the nation, a timid race, and live in con-
tinual dread of the Eskimos, whom they suppose
not only to be very warlike and ferocious, but also
endowed with great conjuring powers, by which
they can compass the death of an enemy at a
distance. The possession of fire-arms does not em-
bolden the Tinnè to risk an open encounter with
the Eskimo bowmen; and unless when they are assembled in large numbers, as we found them at the
Ramparts, they seldom pitch a tent on the banks
of the river, but skulk under the branches of a
tree, cut down so as to appear to have fallen natu-
rally from the brow of the cliff; and they do not
venture to make a smoke, or rear any object that
can be seen from a distance. On the first appearance of a canoe or boat, they hide themselves,
with their wives and children, in the woods, until
they have reconnoitred, and ascertained the character of the object of their fears. More than
once in our descent of the river, when we had
landed to cook breakfast or supper, and were not
at all aware of the vicinity of natives, a family
would crawl from their hiding-places, and come to
our fire. They always pleaded want of food; and
as their wretched appearance spoke strongly of
their necessities, they invariably shared our meals;
but not unfrequently they sold us a fish or two

before we parted; being probably what they had reserved for their next meal, if we had not furnished them with one. We never found them with abundance of food; for, in times of plenty, they do not think it necessary to lay up a stock, but let the future provide for itself.

It is supposed that formerly the Eskimos were in the habit of ascending the river to the Ramparts to collect fragments of flinty slate for lance and arrow-points; but they have been only once so far up, since the trading-posts were established. An old Indian, who was alive within a few years, told Mr. Bell that on that occasion he was wounded by an arrow; but that he succeeded in escaping to the top of the cliff, from whence he killed two Eskimos with his fowling-piece.

As we passed the encampment, the Indians rushed down to the river's side, and, launching their canoes, accompanied us to Fort Good Hope, which now stands near its earliest site, a short way below the defile. At the time of Sir John Franklin's descent of the river in 1825 and 1826, the post stood about one hundred miles further down; but it was removed to its present position in 1836, after the destruction of the former establishment by an overflow of the river. The flood, carrying with it large masses of ice, rose thirty feet; and, mowing down the forest timber, swept onwards to the fort, which it filled with water, thereby destroying a quantity of valuable furs.   Mr. Bell, who was the resident officer at the time, escaped with the other inmates in a boat to the centre of the island; and shortly afterwards, the dam of ice giving way, the flood subsided as rapidly as it had risen, leaving the buildings still standing, though much injured.   A few turnips, radishes, and some other culinary vegetables, grow at Fort Good Hope in a warm corner, under shelter of the stockades; but none of the Cerealia are cultivated there, nor do potatoes repay the labour of planting.   Mr. M'Beath, who had charge of the post, supplied us with some rein-deer venison, which he had kept fresh in his ice-cellar, dug under the floor of his hall.   This gentleman informed us that no rain had fallen this season in his vicinity except two very slight showers on one day: there had been no thunder-showers.   From him we learnt also that a rumour of guns having been heard on the coast of the Arctic Sea, and supposed to have been fired from the Discovery ships, originated in a story brought by the Kutchin or Loucheux to Peel's River Fort, but that the officer in charge placed no reliance upon it.   He also gave us the unpleasant intelligence of three Eskimos having been killed in Peel's River last summer.   A large body of that nation, having ascended the Peel River, it was surmised, with hostile intentions, were fired upon by the Kutchin, and three of them killed, upon which they retreated.

The Kutchin and Eskimos of the estuary of the Mackenzie meet often for purposes of trade, and make truces with each other, but they are mutually suspicious, and their intercourse often ends in bloodshed. The Kutchin have the advantage of fire-arms, but the Eskimos are brave and resolute, and come annually to Separation Point at the head of the delta, for the purposes of barter. Most of the Kutchin speak the Eskimo language, and from them the latter people have become aware of the existence of a post on the Peel. It is probable, therefore, that the Eskimos had a purpose of opening a trade directly with the white people; but this, being so obviously contrary to the interest of the Kutchin, was likely to meet with all the opposition they could offer, and hence their firing on the Eskimos without parley. The Kutchin give a very bad character of their neighbours for treachery, and throw on them the whole blame of their mutual quarrels; but the faults are certainly not confined to one side; and, doubtless, were an intercourse once fairly established between the Eskimos and the Company's posts, it might be kept up as peaceably here as it is with the same people elsewhere.

In the course of Mr. Bell's residence on Peel's River, an event occurred in the history of these people, which, in its principal feature, bore no small resemblance to the skirmish between the parties of Joab and Abner, related in the second chapter of the second book of Kings. A party of Kutchin having met a number of Eskimos, their demeanour to one another was friendly, and the young men of each nation rose up to dance. The Eskimos, however, being accustomed to carry their knives concealed in their wide sleeves, did so on this occasion, and, grasping them suddenly, on a preconcerted signal in the midst of the dance, thrust them at their Kutchin companions, by which three of the latter fell mortally wounded. A mêlée ensued, in which several were slain on both sides. This is the story told by the Kutchin survivors, but the Eskimos would, perhaps, give a different colour to the matter were they the narrators. It is to be hoped that, in a few years, the interference of the traders will put an end to these disastrous conflicts, which have long ceased in other parts of the fur countries.*

* An unexpected and cruel massacre of a party of Eskimos has been reported to the Admiralty by Commander Pullen, since this and some of the following sheets were set in type. This sad occurrence is rendered more lamentable, from a Canadian in the employment of the Hudson's Bay Company having been a prime actor in the affair. It appears that in the spring of 1850,

By Mr. Bell, I was also informed of the melancholy death of an Indian in the vicinity of Fort
Good Hope. This poor man, having set several
snares for bears, went to visit them alone. The
event showed that he had found a large bear, caught
by the head and leg, and endeavoured to kill it with
arrows, several of which he shot into the neck of
the animal. He seems to have been afraid to approach near enough to give full effect to his weapons,
and the enraged bear, having broken the snare, flew
upon him and tore him in pieces. The man's son,
a youth of about sixteen years of age, becoming
alarmed by the lengthened absence of his father,
took his gun, and went in quest of him, following
his track. On approaching the scene of the tragedy,
the bear hastened to attack him also, but was shot

two men belonging to Peel River Fort had landed on Point
Separation, on which a body of Kutchin were at the time
encamped. Soon afterwards a small number of Eskimos approached in their kaiyaks. The Canadian would have fired
upon them instantly, but was restrained by his companion, who
did all that he could to prevent the bloodshed that ensued. The
leading Eskimo called to the Kutchin to lay down their guns ;
and, to show his own peaceful intentions, he fired his arrows into
the sand, and then showed his empty quiver. His signs of amity
were replied to by the Canadian firing upon him and the Kutchin
following his example, the party was destroyed. I fear that our
endeavours to establish friendly relations with the Eskimos in
the estuary of the Mackenzie may have lured these poor people
to the bold advance among their enemies which ended so fatally
for them.

by the lad as he was rushing at him. The boy found his father torn limb from limb, and mostly eaten, except the head, which remained entire. The bear, whose carcass was seen by Mr. Bell, was a brown one, and of great size. Fragments of the snare remained about his neck and leg.

These brown bears are very powerful; and the same gentleman who told the above story informed me that on the Porcupine River, to the west of the Peel, he saw the foot-marks of a large one which having seized a moose-deer in the river, had dragged it about a quarter of a mile along the sandy banks, and afterwards devoured it all, but part of the hind quarters. The bones were crushed and broken by the animal's teeth, and, from their size and hardness, Mr. Bell judged the moose to have been upwards of a year old, when it would weigh as much as an ox of the same age. The species of these northern brown bears is as yet undetermined. They greatly resemble the *Ursus arctos* of the old continent, if they are not actually the same; and are stronger and more carnivorous than the black bears (*Ursus americanus*), which also frequent the Mackenzie. The grisly bears (*Ursus ferox*) reach the same latitudes, but do not generally descend from the mountains.

After a halt of little more than two hours with Mr. M'Beath, we resumed our voyage down the river, and, rowing until supper-time, the crews retired to rest in the boats, which were suffered to drift with the current all night under the guidance of a steersman. On the morning of the 29th, the fog was so dense that for some hours we allowed the boats to follow the current, being afraid to row, lest we should run aground. At night we encamped not far from the Old Fort. The shale, sandstone, and limestone beds, continue throughout the space intervening between the former and present sites of Fort Good Hope. In some places the friable sandstones, yielding readily to the torrents of water which flow over the brow of the cliff in spring, were cut into deep ravines at regular distances, producing conical, truncated eminences, like shot-piles. In others, beds of bituminous shale, one hundred and twenty feet high, existed, interleaved with two or three beds of limestone, and in several places the shale banks were crowned with a thick deposit of sand, which rose above the level of the country behind. This peculiar arrangement, which has been already mentioned as occurring not only on the Mackenzie, but extending also some way up many of its affluents, is conspicuous in the reach immediately above the old site of Fort Good Hope, and has the aspect of ridges of sand left in these situations on the subsidence of waters, that have swept over the neighbouring country. These banks rise much beyond any floods of the present day, some of them being fully two hundred feet above the river. In this neighbourhood, the drift timber showed that the spring accumulations, at the disruption of the ice, occasionally raises the river at least forty feet. Here, as well as higher up, there is generally a capping of diluvium, with boulders, which roll down and line the beach. Among these, sandstones predominate; but there are many of a beautiful porphyritic granite, and others of sienite, hornblendic rocks, greenstone, &c. No clay-slate nor mica-slate boulders were observed.

Vegetation here preserves the same general character that it has higher up the river. *Salix speciosa* continues to grow twenty feet high in favourable localities; the humbler *Salix myrsinites* skirts stony rivulets; and the *Salix longifolia* covers the flooded sandbanks, and arrests the mud. The *Hedysarum boreale* furnishes long flexible roots, which taste sweet like the liquorice, and are much eaten in the spring by the natives, but become woody and lose their juiciness and crispness as the season advances. The root of the hoary, decumbent, and less elegant, but larger-flowered *Hedysarum Mackenzii* is poisonous, and nearly killed an old Indian woman at Fort Simpson, who had mistaken it for that of the preceding species. Fortunately, it proved emetic; and her stomach having rejected all that she had swallowed, she was restored to health, though her recovery was for some time doubtful.* On the beach I observed a patch of parsley (*Apium petroselinum*) in flower ; probably having sprung from seed scattered by a party going to Peel River, as I met with the plant in no other quarter.

July 30th.—In this day's voyage we saw many small parties of Kutchin, seemingly all in want of provisions, owing to the high water spoiling their fishery. From one man, however, we purchased a fine white-fish (*Coregonus*), weighing nearly eight pounds. These families are the most easterly of the Kutchin ; and, far from exhibiting the manly conduct and personal cleanliness for which their nation is noted on the banks of the Porcupine and Yukon, have much of the abject demeanour of their neighbours, the Hare Indians. Their jackets differ from those of the Chepewyans in being peaked, after the manner of those of the Eskimos. From their being able to remain in the close vicinity of the latter people, it is evident that they possess more courage than the Hare Indians.

In the morning we passed an affluent thirty or forty yards wide, coming in from the eastward,

* There must have been some mistake in the information which I furnished to Sir William J. Hooker respecting these two plants, as the *H. Mackenzii* is said in the *Flora Boreali-Americana* to have the edible root.

which is probably the stream mentioned by Sir Alexander Mackenzie as one on whose banks Indians and Eskimos collect flints. These flints are doubtless either chert from the limestone beds, or flinty slate, which exists plentifully in some parts of the shale formation.

Early in the morning of the 31st, we ran through the " Narrows," a defile similar to that of the Ramparts, and in passing which the river makes a similar sharp elbow. The cliffs are composed of sandstone, in some places horizontal, in others dipping to the south by east at a small angle. The stone is of various textures; some of it having a conchoidal fracture, and containing much calcareous matter. The basis is earthy, and the coarsest stone is composed of small, rounded, and also sharply angular grains of opake, white, green, or blue quartz, with grains of Lydian-stone and coal; the basis being also tinged with coaly matter. Other beds pass into a kind of wacké, or shale, which breaks down quickly into very small angular fragments. This shale is often encrusted with alum in powder, and it is sometimes stained with iron, and contains spheroidal nodules of clay-ironstone. The cliffs vary in height from fifty to one hundred and fifty feet, and the capping of clay and loam with boulders is thin. The shale formation extends along the banks of Peel's River; and Mr.

Bell informed me that he had procured crystallised alum from some beds in that quarter. The Mackenzie is about one mile and a half wide in these straits. The current sets at the rate of three miles an hour, and we got soundings with eight fathoms in mid-channel. By a mutual understanding between the Eskimos and Kutchin, the Red River, which falls into the Narrows, is considered as the boundary between the two nations.

On emerging from the Narrows, we had a distant view of the Richardson chain of hills, which skirts the western branch of the Mackenzie, and a little before noon reached Point Separation. Mr. Rae observed for the latitude; but the sun being obscured by passing clouds, the observation would have been doubtful, had it not corresponded exactly with the position assigned to the spot in Sir John Franklin's map of 67° 49′ north latitude. The variation of the compass by the sun's meridional bearing was south 55° east, being five degrees more than in 1826.

In compliance with my instructions, a case of pemican was buried at this place. We dug the pit at the distance of ten feet from the best-grown tree on the point, and placed in it, along with the pemican, a bottle containing a memorandum of the objects of the Expedition, and such information respecting the Company's post as I judged would be useful to the boat party of the " Plover," should they reach this river.   The lower branches of the tree were lopped off, a part of its trunk denuded of bark, and a broad arrow painted thereon with red paint. A stake was also erected on the beach, and a paper attached thereto, directing attention to the tree.   We considered it likely that the stake might be taken down by Indians or Eskimos; but as the latter people lop trees in the same manner, we did not think the circumstance of one being so cut here would induce people of either nation to dig in the vicinity.   To conceal as effectually as we could that the earth had been moved, the soil was placed on a tarpaulin, and all that was not required to fill the hole was carried to a distance.   The place then being smoothed down, a fire of drift timber was made over it, that the burnt wood might indicate the exact spot to the " Plover's " party, who were furnished with a memorandum mentioning that these precautions would be used.   Along with the pemican, a letter for the purser of the " Herald," which his friends had committed to my care, was placed in the pit.*

* Commander Pullen, with two boats from the " Plover," in Sept. 1849, visited this depôt, and found it safe. The lopped tree had previously, in autumn 1848, been examined by a party of the Hudson's Bay Company's servants going to Peel River, but they did not discover the pit, having no key to enable them to find it.

In performing these duties at this place, I could not but recall to mind the evening of July 3rd, 1826, passed on the very same spot in company with Sir John Franklin, Sir George Back, and Lieutenant Kendall. We were then full of joyous anticipation of the discoveries that lay in our several paths, and our crews were elated with the hope of making their fortunes by the parliamentary reward promised to those who should navigate the Arctic Seas up to certain meridians. When we pushed off from the beach on the morning of the 4th to follow our separate routes, we cheered each other with hearty good will and no misgivings. Sir John's voyage fell some miles short of the parliamentary distance, and he made no claim. My party accomplished the whole space between the assigned meridians; but the authorities decided that the reward was not meant for *boats*, but for ships. Neither men nor officers made their fortunes; and, what I more regretted, my friend and companion, Lieutenant Kendall, remained in that rank till the day of his death, notwithstanding his subsequent important scientific services. On the present occasion, I endeavoured to stimulate our crews to an active look-out by promising ten pounds to the first man who should announce the Discovery ships.

Most of the islands constituting the delta of the Mackenzie are alluvial, and many of the smaller ones are merely a ring of white spruce trees and willows on a sand or mud bank, enclosing ponds or marshes filled with drift timber. Some of the larger ones have a drier and firmer soil, but are low and even, except near the sea, where a few conical hummocks rise abruptly above the general level to the height of eighty or ninety feet. Sir John Franklin saw these hummocks on Ellice Island ; and, as they occur also near the western boundary of the delta, I shall have occasion to notice them again. The Richardson Mountains, which skirt the western channel of the river, appear like a continuous ridge when viewed from Point Separation ; but it is more probable that they are the termination of a succession of spurs from the main chain of the Rocky Mountains, which come obliquely to the coast of the estuary. The general altitude of the ridges does not apparently exceed one thousand feet ; but some peaks, as Mount Goodenough, Mount Gifford, and Mount Fitton, are perhaps considerably higher.

In ascending from the west bank of the delta, the brow of the first range is attained at the distance of about forty miles from the river, of which the first four miles are over a low, marshy, alluvial plain, covered with willows. Two or three almost precipitous ascents and descents across other mountains bring the traveller to a small stream, called the Rat River, which flows to the westward. This is said to issue from a lake, which also gives origin to a still smaller stream, bearing likewise the appellation of the Rat, and taking an opposite course to join the western branch of the Mackenzie. The Western Rat River is an affluent of a considerable stream, named the Porcupine, which, running to the west-south-west for two hundred and thirty miles, enters the Yukon, a river emulating the Mackenzie in size, and flowing parallel to it, but on the western side of the Rocky Mountains. Of the country watered by this great river, and its inhabitants, I shall take occasion to speak hereafter.

The eastern channel of the delta of the Mackenzie is also flanked by a ridge, named the Reindeer Hills, which I consider to be a prolongation of the spur that the Mackenzie crosses at the " Narrows." They are not so rugged or peaked in their outline as the Richardson chain, or the spurs which the Mackenzie passes through higher up ; and their general height does not appear to exceed seven or eight hundred feet.

Having finished the operations at the *cache*, we resumed the voyage, and, retracing our way for a few miles, entered the eastern channel of the delta, and pursued it until seven in the evening, when we encamped, about twenty-two miles below Point Separation. The banks of the river here, and the numerous islands, are well wooded. The balsam poplars rise to the height of twenty feet, and the white spruces to forty or fifty. Numbers of sand martens burrow in the banks. These birds winter in Florida. Mr. Audubon informs us, that in Louisiana they begin to breed in March, and rear two or three broods in a season. In the middle states their breeding-time commences a month later; and in Newfoundland and Labrador it rarely takes place before the beginning of June. Near the mouth of the Mackenzie, the banks are scarcely thawed enough to admit of excavation by the feeble instruments of this bird before the end of June; and in the beginning of September, the frosts prostrating the insects on which the martens feed, they and their young broods must wing their way southwards. I was unable to procure a specimen of this marten, though it breeds in multitudes along the whole course of the Mackenzie, and am therefore unable to decide whether it is the *Hirundo riparia* or *Hirundo serripennis* of Audubon; but from its nearly even tail, I rather incline to think it may be the latter; and if so, it may not be the same species which breeds in the southern states. The sand marten was first seen by us on the 28th of May, as we were descending the River Winipeg, near the 50th parallel, and we know, from our observations in 1826, that it reaches the delta of the Mackenzie by the beginning of July; affording thus an index to the progress of spring in different latitudes. On the Winipeg it was accompanied by the purple swift (*Progne purpurea*), whose northern limit we did not ascertain.

We resumed our voyage at three in the morning on the 1st of August, and when we landed to cook breakfast, saw some recent footmarks of Eskimos. As these people are employed at this time of the year, in hunting the rein-deer on the hills which we were skirting, we were in constant expectation of seeing some of their parties. The Rein-deer Hills, as viewed from the eastern channel, seem to be an even-backed range; but when examined with the telescope, they are seen to consist of many small, oblong, rocky eminences, apparently of limestone, and are sparingly wooded. In the course of the morning we crossed the mouths of three pretty large affluents, coming in from the hills, and also two cross canals, dividing M'Gillivray Island into three sections.

About thirty-five miles from Point Separation, or in latitude 68° 10′ N., the channel washes the foot of a low dome-shaped bluff, in which the intrusion

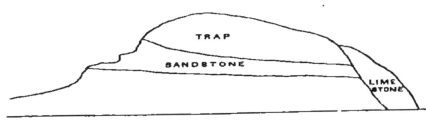

of a mass of trap, which now forms the top of the hill, has tilted up a bed of limestone, and separated it from one of sandstone.

In the afternoon we passed another considerable affluent from the hills in lat. 68° 18′ N.; some hours later, another one of less size; and very soon afterwards crossed a channel which bounds Harrison Island on the south. This island, like M'Gillivray's, is divided into several portions by minor creeks. The boats were under sail all the afternoon, and must have been observed, about 5 o'clock, by the hunting-parties of the Eskimos, for at that time we noticed a line of six or eight signal smokes, raised in succession along the hills, and speedily extinguished again. As the Eskimos use fire-wood very sparingly for cooking, and, like the Indians generally, burn only dry wood which emits but little vapour, we knew that the smokes we saw were intended to spread the intelligence of the arrival of strangers in the country, and therefore that we might expect to find a considerable body assembled on some part of the river to meet us. In the evening we landed to cook supper, and re-embarked to continue under sail all night, with a very light breeze; our progress was, however, slow, owing to the uncertain eddies and currents, produced by the junction of the several cross-channels. At midnight we passed the creek which bounds Harrison Island on the north, in 68° 37′ N. Here several gently swelling elevations interpose between the river and the main ridge of the Rein-deer Hills. The valleys and borders of the river are well wooded, but the summits of the eminences present only scattered spruce firs, with stunted tops and widely spreading depressed lower branches. The canoe-birch (*Betula papyracea*) is frequent, and the trees we measured were about five inches in diameter. The *Populus balsamifera* and *Alnus viridis* grow to the height of twenty feet, and the *Salix speciosa* to upwards of twelve. The *Ribes rubrum*, *Rubus chamæmorus*, and *Vaccinium vitis idæa*, bore at this time ripe fruit. The *Rosa blanda*, *Kalmia glauca*, *Nardosmia palmata*, and *Lupinus perennis*, were also observed flourishing in this high latitude, together with several other plants which extend to the sea-coast. Among the birds, we saw the great tern (*Sterna cayana*), the *Coryle alcyon*, and *Scolecophagus ferrugineus*, the latter in flocks.

August 2d.—For five or six hours this morning we ran past the ends of successive ridges separated by narrow valleys. The diagram gives the outlines of one of these spurs seen on the southern flank. It is about three hundred feet high, and its acclivities are furrowed deeply, producing conical eminences which are impressed with minor furrows. The vegetation is scanty; a few small white spruces straggle up the sides; and the soil, where it is

Sand hills.   Lat. 65° 50′ N.

exposed to view, is a fine white sand.   Large boulders lie on the sides of the hills, and, judging
from the structure of the only point on which
time permitted me to land, the whole appears to
be similar to the sand deposit with its capping of
boulder gravel which covers the shale on Athabasca
and Mackenzie Rivers.   On the point in question,
the white sandy soil was ascertained to come from
the disintegration of a sandstone, which has just
coherence enough when *in situ* to form a perpendicular bank, but crumbles on being handled.   It
consists of quartz of various colours, with grains of
Lydian-stone loosely aggregated, and having the
interstices filled with a powdery matter, like the
deposits of some calcareous springs.   Similar sandstones occur at the "Narrows."   Above it, there is
a bed of gravel, also formed of variously coloured

grains of quartz, mixed with chert from limestone. Most of the quartz is opaque, and veined or banded, but some of it is translucent. Some bits are bluish, others black, and many pebbles are coloured of various shades of mountain green. The latter are collected by the Eskimos and worn by them as labrets. The gravel covers the whole slope of the point, which is so steep as to require to be ascended on all fours. In one part a torrent had made a section of a bed of fine brown sand, twenty feet deep. On this bank I gathered the *Bupleurum ranunculoides*, which grows in Beering's Straits, but had not been found so far westward as the Mackenzie before; also the *Seseli divaricatum*, which had not been previously collected to the north of the Saskatchewan.

In latitude 68° 55′ N. the trees disappeared so suddenly, that I could not but attribute their cessation to the influence of the sea-air. Beyond this line a few stunted spruces, only, were seen struggling for existence, and some scrubby canoe-birches, clinging to the bases of the hills. Further on, the Rein-deer Hills lowered rapidly, and we soon afterwards came to Sacred Island, which, with the islets beyond it, is evidently a continuation of the sandy deposit noticed above. Had time permitted, I should have gone past Sacred Island, northwards, to deposit some pemican on Whale Island, but at so advanced a period of the summer,

I was unwilling to incur the loss of a day which that route was certain to occasion, and perhaps even of two days.

We did not land on Sacred Island, but observed in passing that it still continued to be a burying-place of the Eskimos; two graves covered by the sledges of the deceased, and not of many years' construction, being visible from the boats.   This is the most northerly locality in which the common red currant grows on this continent, as far as I have been able to ascertain.   Five miles beyond the island, we landed on the main shore, to obtain a meridional observation, by which the latitude was ascertained to be 69° 4′ 14″ N., and the sun's bearing at noon, south 51° east.   About three miles further on, we had a distant view of an eminence lying to the east-

Conical Hill behind Point Encounter.

ward, which resembled an artificial barrow, having
a conical form, with very steep sides and a truncated summit. This summit, in some points of
view, presented three small points, in others, only
two, divided from one another by an acute notch.
In the afternoon I landed on Richard's Island, which
rises about one hundred and fifty or two hundred
feet above the water, has an undulated grassy
surface, and is bordered by clayey or sandy cliffs
and shelving beaches. The main shore has a
similar character. The channel varies in depth
from two to six fathoms, but is full of sand-banks,
on which the boats frequently grounded.

At ten in the evening we encamped on Point
Encounter, in latitude 69° 16′ N., and set a watch
at the boats, and also on the top of the bank, which
is here nearly two hundred feet high. The tide
ebbed at the encampment, from seven in the evening
till half an hour after midnight. The ensign was
planted on the summit of the cliff all the evening,
and was no doubt seen by the Eskimos, who were in
our neighbourhood, and most probably reconnoitred
our encampment, but we saw nothing of them.

The readers of the narrative of Sir John Franklin's Second Overland Journey will recollect that
off this point the Eskimos made a fruitless attempt
to drag the boats of the eastern detachment on
shore, for the purpose of plundering them.

CHAP. VIII.

August 3rd, 1848.—HAVING given some verbal
instructions to the crews of the boats, respecting
their conduct in the presence of the Eskimos, we
embarked at four in the morning, and, crossing a
shallow bar at the east end of a sand-bank, stood
through the estuary between Richard's Island and
the main, with a moderate easterly breeze, which
carried us gradually away from the main shore.
About an hour after starting, we perceived about
two hundred Eskimos coming off in their kaiyaks,
carrying one man each, and three umiaks filled
with women and old men, eight or ten in each.
The kaiyaks are so easily overbalanced, that the

sitter requires to steady it before he can use his bow or throw his spear with advantage, unless when three lie alongside each other, and lay their paddles across, by which the central man is left at liberty to use both hands. By taking the precaution, therefore, of not allowing the Eskimos to hamper us, by clinging to the boats, and continuing to make some way through the water with the oars, we were pretty sure that they could not take us altogether by surprise; and I felt confident that as long as they saw that we were on our guard, and prepared to resist any aggression, none would be attempted. I had, moreover, especially directed Duncan Clark, who was cockswain of the third boat, in which there was no other officer, to keep close to mine, which he could easily do as his was the swifter of the two ; but the novelty of the scene caused him to neglect this command, fortunately with no serious bad consequence, though a conflict might have been the result of his inattention to orders.

Mr. Rae and I carried on a barter with the men in the kaiyaks, paying them very liberally for anything they had to offer in exchange, such as arrows, bows, knives of copper or of bone, &c., and thereby furnishing them with much iron-work, in the shape of knives, files, hatchets, awls, needles, &c. The articles we received were of no value to us; but a gift is generally considered by the American nations as an acknowledgment of inferiority, and it is better to exact something in exchange for any article that you may wish to bestow. The men were very persevering in their attempts to hold on by the boats, and we were obliged to strike them severely on the hands to make them desist. Previous experience had taught me the absolute necessity of firmness in repressing this practice, and I was pleased as well as surprised to see the patience with which they generally endured this treatment, — a few only of the bolder spirits showing a momentary anger, but all acquiescing at length in the rule we had laid down. The freshness of the breeze which blew during our intercourse rendered it easier to deal with them, as they dropped behind directly they had ceased to ply their paddles; but they had no difficulty in out-stripping our boats whenever they exerted themselves; and I have little doubt but that they are able to propel their light kaiyaks at the rate of seven miles an hour.

While we were thus engaged, we heard the report of two muskets from the third boat, which had dropped two or three hundred yards astern. It appeared that the kaiyaks had not been so rigidly kept off by Clark, but had been allowed to hamper the oars so as to retard the boat's way. Some of the Eskimos, paddling close up to the stern, had tried to drag the cockswain overboard by the skirts of his jacket; and performed other various aggressive acts, evidently with the view of ascertaining to what lengths they might proceed with impunity. The umiaks, which had been kept aloof from the foremost boats, made a push for the third one, and one of them running across her bows, the men and women it contained instantly began to plunder the boat, and to struggle with the crew, who, being only six in number, would have been soon overpowered. Immediately on hearing the reports of the muskets, which were fired in the air merely as signals, I wore my boat round amid the shouts of the Eskimos who were hovering near us, and who thought that I was about to comply with their urgent requests that we should land and encamp in their neighbourhood. Hauling up under the stern of Clark's boat, I declared that I would immediately fire upon the assailants if they did not desist, and my crew at the same time presenting their muskets, the attempt was at once quelled. I found, on subsequent inquiry, that nothing had been carried away but a small box on which the cockswain sat, which contained his shaving apparatus, and some other trifling articles belonging to himself, together with the boat's ensign. Though vexed at the loss of the latter, I could not but be glad that he was the only one of the crew who suffered, and I looked upon the theft of his property as poetical justice for his want of firmness. The umiaks, not being suffered again to approach our boats, soon pulled towards the shore, but the majority of the kaiyaks continued to keep company with us, the men conversing and begging as if nothing unpleasant had occurred. At length, as we had been drawing away from the shore all the morning, we totally lost sight of the land ; and the kaiyaks assembling together, the men held a short consultation, and then paddled towards their encampment, being guided in their course by a dense column of smoke, which their families on shore had raised. Four or five of them continued to follow us for a short time, after the great body had gone away, evincing their boldness, even when much inferior in numbers, but they, also, went off on receiving some presents, which we could then make to them without fear of misconstruction.

Our inquiries were directed chiefly to obtaining information of the Discovery Ships, but the Eskimos, one and all, denied having ever seen any white people, or heard of any vessels having been on their coast. None acknowledged having been present at the various interviews of their countrymen with white people in 1826, and perhaps the circumstances attending those meetings might have deterred them from confessing that they were relatives of the parties that assailed Sir John

Franklin's boats at that time; and as most of the men were stout young fellows, and few beyond the prime of life, only two or three of the old men in the umiaks could have been actually engaged in the struggle which then took place. One fellow alone, in answer to my inquiries after white men, said, " A party of men are living on that island," pointing, as he spoke, to Richard's Island. As I had actually landed there on the preceding day, I was aware of the falsehood he was uttering; and his object was clearly to induce us to put about and go on shore, which he and others had been soliciting us to do from the commencement of our conversation. I, therefore, desired Albert to inform him, that I had been there, and knew that he was lying. He received this retort with a smile, and without the slightest discomposure, but did not repeat his assertion. Neither the Eskimos, nor the Dog-rib or Hare Indians, feel the least shame in being detected in falsehood, and invariably practise it, if they think that they can thereby gain any of their petty ends. Even in their familiar intercourse with each other, the Indians seldom tell the truth in the first instance; and if they succeed in exciting admiration or astonishment, their invention runs on without check. From the manner of the speaker, rather than by his words, is his truth or falsehood inferred; and often a very long interrogation is necessary to elicit the real fact. The comfort, and not unfrequently the lives, of parties of the timid Slave or Hare Indians are sacrificed by this miserable propensity. Thus, a young fellow often originates a story of his having discovered traces of an enemy for which there is no real foundation. This tale, though not credited at first, makes some impression on the fears of the others, and soon receives confirmation from their excited imaginations. The story increases in importance, a panic seizes the whole party, they fly with precipitation from their hunting-grounds, and if they are distant from a trading post, or large body of their nation, many of the number often perish in their flight by famine.

The Eskimos are essentially a littoral people, and inhabit nearly five thousand miles of sea-board, from the Straits of Belleisle to the Peninsula of Alaska ; not taking into the measurement the various indentations of the coast-line, nor including west and east Greenland, in which latter locality they make their nearest approach to the western coasts of the old world. Throughout the great linear range here indicated, there is no material change in their language, nor any variation beyond what would be esteemed in England a mere provincialism. Albert, who was born on the East Main, or western shore of James's Bay, had no

great difficulty in understanding and making himself understood by the Eskimos of the estuary of
the Mackenzie, though by the nearest coast-line
the distance between the two localities is at least
two thousand five hundred miles. Traces of their
encampments have been discovered as far north
in the new world as Europeans have hitherto
penetrated; and their capability of inhabiting these
hyperborean regions is essentially owing to their
consuming blubber for food and fuel, and their
invention of the use of ice and snow as building
materials. Though they employ drift-timber when
it is available, they can do without it, and can
supply its place in the formation of their weapons,
sledges, and boat-frames, wholly by the teeth and
bones of whales, morses, and other sea animals.
The habit of associating in numbers for the chase
of the whale has sown among them the elements
of civilisation; and such of them as have been
taken into the Company's service at the fur posts
fall readily into the ways of their white asso-
ciates, and are more industrious, handy, and intelligent than the Indians. The few interpreters
of the nation that I have been acquainted with
(four in all) were strictly honest, and adhered
rigidly to the truth; and I have every reason
to believe that within their own community the
rights of property are held in great respect,

even the hunting-grounds of families being kept sacred. Yet their covetousness of the property of strangers and their dexterity in thieving are remarkable, and they seem to have most of the vices as well as the virtues of the Norwegian Vikings. Their personal bravery is conspicuous, and they are the only native nation on the North American continent who oppose their enemies face to face in open fight. Instead of flying, like the Northern Indians, on the sight of a stranger, they did not scruple in parties of two or three to come off to our boats and enter into barter, and never on any occasion showed the least disposition to yield any thing belonging to them through fear.

As the narratives of the recent arctic voyages contain descriptions of the manners, customs, and features of these people, and the sketches of Captains Beechey, Lyons, Sir George Back, Lieutenant Kendall, and others, give correct delineations of the personal appearance and costume, I shall not say more of them in this place.

As soon as the last of the kaiyaks disappeared beyond our horizon, we struck the boats' masts, and, pulling obliquely in towards the shore, landed to cook our supper at a place where there were three winter Eskimo habitations, and which is situated about seven or eight miles to the eastward of Point Warren. These buildings are generally placed on points where the water is deep enough for a boat to come to the beach, such a locality being probably selected by the natives to enable them to tow a whale or seal more closely to the place where it is to be cut up. The knowledge of this fact induced us generally to look for the buildings when we wished to land. The houses are constructed of drift-timber strongly built together and covered with earth to the thickness of from one to two feet. Light and air are admitted by a low door at one end; and even this entrance is closed by a slab of snow in the winter time, when their lamps supply them with heat as well as light. Ten or twelve people may seat themselves in the area of one of these houses, though not comfortably; and in the winter the imperfect admission of fresh air and the effluvia arising from the greasy bodies of a whole family must render them most disagreeable as well as unwholesome abodes. I have been told that when the family alone are present, the several members of it sit partly or even wholly naked.

As soon as supper was prepared, we withdrew to the boats to eat it, having anchored them under a sand-bank about a bow-shot from the shore. We had scarcely assumed this position when a party of Eskimos were perceived coming round a point a little to the westward, crouching under the bank, and evidently hoping to surprise us at the fire, of which they had seen the smoke. As soon as they knew they were perceived they brandished their bows and knives, made gestures of defiance, and threw their bodies into various most extraordinary attitudes, as they are accustomed to do when they meet with a strange party, of whose intentions they are uncertain.

As we were all, both men and officers, exhausted by the constant clamour, watchfulness, and exertion sustained throughout the morning, and I wished the men to have some repose, we determined on having no communication with this warlike group, since we could hope for no additional information from people residing so near the large body we had so recently parted from. On coming to the winter-houses, they showed themselves, at first cautiously, then openly, and ceased to gesticulate. Soon after-wards one began to wade out to the sand-bank, and another to launch a kaiyak which he had brought. Thereupon, to intimate that we declined to receive them, Mr. Rae fired a ball so as to strike the water a few feet on one side of him. The bold fellow leaped nto the air, cut several capers of defiance, and then retreated with the others into one of the houses, where they thought themselves safe, and from whence they continued to watch us. After a time

one of them waded out to the sand-bank, planted a
stick on it, and, pulling off his rein-deer jacket,
hung it up, intimating that he wished to have its
value, and then retreated to the shore. We, however,
declined bartering, and at 10 P. M. weighed anchor,
and, standing out to sea, worked to windward all
night against a stiff breeze.

August 4th.—We gained only a few miles by
the night's operations, having to contend with a
tumbling sea, which drove us to leeward; and at
3 in the morning, on the wind moderating sufficiently to allow us to use the oars, we struck the
masts and pulled for three hours across Copland
Hutchison Inlet, when we landed on its eastern
side to prepare breakfast. The shore in this
quarter is for the most part low, but, at intervals of
seven or eight miles or more, some of the conical eminences already mentioned occur. They have not
the ridged and escarped aspect of sand hills, but,
on account of their isolation, look more like artificial barrows, though unquestionably they cannot be
works of art. I am inclined to think that they are
remnants of the sand formation which covers the
shale so extensively along the banks of the Mackenzie, and that they have received their conical
form from the washing of high tides during the
occasional inundations of the low-lands by the sea.

Copland Hutchison Inlet is about ten miles

across, and its mouth is obstructed by sand-banks. A river seems to flow into it, as we could trace sandy cliffs for some distance inland, like the banks of a stream. Though many small ponds existed at the place where we landed, they were mostly brackish, and we had to search for some time before we obtained fresh water. On, at length, discovering some, we filled our breakers, to avoid a similar detention at our next meal. Drift-wood was also scarce. From these causes we were unusually long at breakfast; and soon after embarking we landed again to observe the meridional altitude of the sun; by which the latitude was ascertained to be 69° 44′ N., and the variation of the needle 58° E.

As we advanced to the eastward in the afternoon the coast became still flatter, so that the beach was under the horizon when we were in no more than half a fathom of water. The hummocks above mentioned, which came into view in succession, looming like conical islands, as we ran along with a light westerly breeze, were the only land in sight. In blowing weather, the only resource for boats on this coast is to keep a good offing, as the surf then breaks high on the shelving flats. During the afternoon the sky was lurid, as if loaded with fog, and, though the horizon was tolerably clear, objects were very much altered by mirage. Altogether I have never, either in this climate or in any other, seen a more disagreeable atmosphere, and we all predicted a coming storm. At night, the wind heading us, so that we could not fetch the eastern-most hummock in sight, we ran under a sand-bank, and anchored in a foot and a half of water about a quarter of a mile from the shore. On the tide ebbing, the boats were left dry.

The perfectly flat land here is covered, to the depth of four or five feet, with a moorish or peaty soil, which is much cracked, and in many places treacherously soft and boggy. Small lakes and ponds intersect it in all directions, mostly filled with brackish water, but some of them containing water fresh enough for cooking, though by no means good. The irregular ponds and marshy places make the course to any particular point exceedingly devious, and I had a long walk to reach one of the eminences, though its direct distance from the beach was little more than a mile. This hill rose from the boggy ground, in a conical form, to the height of about one hundred feet, its base having a diameter about equal to its height. A ditch, fifteen or twenty feet wide, which surrounded it, was passable only at two points; and, on ascending the hill, I found that hollow, like the crater of a volcano; in which, about fifteen feet below its brim, stood an apparently deep lake of very clear and sweet water. The interior beach of this curious pond was formed of fine clean gravel, and the hill itself apparently consisted of sand and gravel with a coating of earth. From its summit, I saw many similar heights to the eastward and south-eastward, that is, in an inland direction, and they seemed to rise, like islands, out of a great inlet or bay, for the low land connecting them, if such existed, was not visible. I looked long on the scene, but could not satisfy myself whether what I beheld was actually water, or merely the mirage of a low fog simulating an inlet of the sea. I was inclined to consider that the latter was the real state of the case, since, during the whole afternoon, the hills we had passed, as well as the one on which I then stood, seemed equally to rise out of a hazy sea, when we were at the distance of only four or five miles from them. Could I have convinced myself that the isolated peaks I saw were really islands, I should have thought that I was on the borders of the extensive Eskimo Lake, laid down as problematical in Sir John Franklin's chart, but which I now believe does not exist. The Indians, upon whose report it was indicated, meant, most probably, by the expressions they used, the sea itself, or perhaps the inlet of Liverpool Bay, which lies further to the westward. Round the lower part of the hill, about eight or ten feet above the then

level of the sea, lay a ring of large drift-timber,
showing that, in certain states of the wind, the sea
rises so as to inundate all the low lands.   Nothing
that we observed in the whole course of our
voyage led us to think that the spring tides rise
more than three feet on any part of this coast, and
their rise was more generally only about twenty
inches.

Rein-deer frequent these flat lands at this season
to feed on bents and grasses, and the ponds are
full of geese and ducks.   Many red-throated divers
(*Colymbus septentrionalis*) also resort thither, and
utter a most mournful cry when any one invades
their retirement; the Lapland finch (*Plectrophanes
lapponicus*) and horned lark (*Phileremos cornutus*)
make a breeding station of this coast, and we
observed the young birds running in numbers over
the ground.

While Mr. Rae was out hunting, the crews were
cooking supper on shore, and I remained with two
boat-keepers at the boats, I received a visit from
two middle-aged Eskimos and their wives; the
latter being fat, jolly-looking dames, and considerably younger than their husbands.   Albert was
absent, so that I could not profit by the interview to ask as many questions as I desired to do,
but I made out that if we kept in shore of the
sand-banks, we should have water enough for our

boats when the tide rose, and would find a passage out to sea at a point on which their winter
houses stood, — or, as they invariably termed such
stations in their own language, Iglulik.   They
had not seen any white men on their coast, but
with a ready flattery expressed their affection for
them.   Having been very liberal to my visitors,
not only in purchasing all the small articles of
traffic they had to offer, but also by making them
some useful presents of files, hatchets, and knives,
with a considerable quantity of beads, I became
tired of their company and of the constant vigi-
lance required to prevent them from pilfering, and
therefore requested that they would go away,
since I would give them nothing more.   Their
smiling countenances and deferential manner prevented me from using any threat, but I had to repeat my request several times, and, at length, from
the urgency with which I spoke, one of the women
seemed to think that I was afraid of their using
violence, and opening the hood of her jacket, she
showed me that there was a little naked infant in
it.   She then explained that when they went to
war they left the children behind, and that she and
the other women had brought their young ones as
a pledge of their pacific intentions.   I prevailed
upon them at length to go, by assuring them that
I would part with nothing more at that time, but

would give them something on a future occasion, if they came again to us.

Mr. Rae went in pursuit of rein-deer in the afternoon, but the talking of the men who were wandering about scared away the animals. He was more successful in the night, and, by taking Halkett's portable boat on shore with him, killed nineteen brent geese, and some ducks. The water-fowl at this time were moulting in the small lakes, and became an easy prey.

Aug. 5th. — A strong gale of wind raised a high surf on the shoals under which we lay, so that we could not launch the boats over them sea-wards. We, therefore, made sail when the tide rose, and ran on the inside of them until we became involved among the flats, and the boats were left quite dry. During low water, we marked off the best channels with poles. The tide began to flow at $11\frac{1}{2}$ A.M., and by $2\frac{1}{2}$ P.M., there was water enough for us to proceed, by launching the boats over some bars which intervened between the place where we were detained, and deeper water. Following a narrow channel till 5 in the afternoon, we secured the boats in a snug harbour under a tongue of sand at Point Atkinson. A strong north-west gale had raised a high sea, and we could do no otherwise than remain here at anchor until the wind and sea subsided.

Point Atkinson is a flat, low piece of ground, with a range of sand hills, forty or fifty feet high, thrown up along its northern side by the winds and waves. When we visited it in 1825, its extreme point was a small island separated from the main by a ditch but this was now choaked up and formed a marshy pond, the water of which being brackish, and fetid as well as greasy, from the quantity of whale oil with which the ground was saturated, was totally unfit for use. The oil had acted as a manure on the soil, and produced a luxuriant crop of grass from one to two feet high (*Elymus mollis*, *Calamagrostis stricta*, *Spartina cynosuroides*, and some shorter *Carices*.) A small village of seven or eight huts stood on the point, among which is the *Kashim*, or house of assembly, described and figured in the narrative of my former voyage along this coast. I am not aware that a house of this description, appropriated as a council chamber and eating-place for the males, has been described as existing among the more eastern Eskimos. They possess, however, the appellation for it in their language. It is of more importance among the western tribes, as I shall have occasion to mention in a subsequent chapter.

The sea has carried away much of the sandy bank on which the *Kashim* stood, and the spray now washes its walls, so that it will likely be over-

thrown in a few seasons. It is still supported by
whale-skulls built round its outside wall. After
seeing the boats secured, Mr. Rae and I walked
through the village by a well-beaten path which
meandered among the long grass; and on looking
into one of the smaller houses, found an old crone
sitting warming herself over a few embers. She
seemed surprised when she saw our faces, but
exhibited no signs of fear, and soon began to talk
very volubly. Some rows of very inviting herring-
salmon being hung to dry on poles by her tent,
she gave me five or six at my request, and, while
Mr. Rae had gone to the boats to procure some
articles to repay her and for presents, she told me
that she must go to a party who were encamped
on the shore about a mile off. I endeavoured to
persuade her to remain till she got the articles I
had sent for, but she was bent on going, and,
having made up a small bundle of some of her
property, walked off with a quickness and elasticity
of step which her first appearance was far from war-
ranting. The deep furrows of her countenance, and
its weather-beaten aspect, gave her the appearance
of advanced age; and while she sat she seemed to be a
woman of ordinary stature, a little bowed down; but
when she stood up she looked dwarfish, her height
not exceeding four feet. Two tents, towards which
she bent her steps, were visible from the boats, but

we thought it better to let the inmates recover from their surprise at our arrival, and to seek an interview in their own way.

After a time we saw a man coming towards us. He made his approach very slowly, and by a devious path, keeping a muddy channel between us and him. On Mr. Rae going alone and unarmed to meet him, he waited his approach with evident trepidation, and, when they met, began to express by signs and words that he was old, infirm, and nearly blind, which was by no means the truth. Mr. Rae invited him to come to our tents, which were by this time pitched, and, after much persuasion, induced him to come on, but not until he had, on the request of the old man, blown in each of his ears, tapped his breast and touched his eyes, as a charm, either to remove his maladies, or more probably to avert any evil influence which the white men might possess. After the performance of these ceremonies he came to us; and as his confidence increased he gradually laid aside the appearance of infirmity, and began to bustle about and pry into every thing, until at length he became troublesome. Though repeatedly spoken to, and told that we would not suffer him to handle any thing belonging to us, he was scarcely restrained, and required constant watching to prevent him from stealing. He took up his abode in one of the huts, and, after we had retired to rest, made an attempt to raise one of the boat's anchors, and continued to prowl about until the sentinel on duty checked him by showing his musket. In the middle of the night he entered one of the tents, with a long knife in his hand, but retreated on being spoken to. Perhaps he had no intention of committing any violence, as he habitually carried this knife, which was of Russian manufacture, either in his hand or up his sleeve. Through Albert I learnt that the two natives whom we had seen on the preceding day were sons of this man, and that they, with their families and some young people who were hunting with them, would come to the village in a few days to engage in the chase of whales, when they would be joined by one or two parties then in the pursuit of game on the east side of a river which falls into M'Kinley Bay. This small community does not wander far from their winter station on Point Atkinson. The hunters pursue rein-deer and water-fowl on the neighbouring flats in summer, chase the whale during one month or six weeks of autumn, live with their families in the village during the dark winter months, and in spring travel seaward on the ice to kill seals, at which time they dwell in snow-houses. The old man and several of his elderly companions who subsequently came to the village declared that they had no remembrance of my former visit to this coast, and

said that now was the first time that they had
seen white men. They do not go as far as the
mouth of the Mackenzie, and dread their turbulent
countrymen in that quarter.

As none of the ponds or ditches which intersect
Point Atkinson in all directions contain drinkable
water, we were on first landing at some loss. The
old woman, however, had told us that her people
procured water from the seaward side of the sand-
hills, and, by following a path which led in that
direction, we discovered three wells, carefully built
round with drift-timber, below high-water mark;
which, when we first saw them, were completely
sanded up. On clearing them out water contaminated with fetid whale oil flowed in abundantly, but
this being repeatedly drawn off, until the surrounding sand was washed from its impurities, we
at length obtained tolerable water for making tea.
These wells are evidently supplied from rain falling
on the sand-hills, and kept up to the level at which
we found it by the pressure of the sea.

August 6th.—The old woman whom we first
saw, another still older, and an aged blind man,
came to the village this day; and in the afternoon
three fine young men brought some ducks, which
we purchased from them. They were eager to sell
water-fowl for buttons, beads, or any trifle we chose
to offer, and our crews eventually obtained a con-

siderable number in that way; but they were very unwilling to dispose of the fish which hung on poles in the village. After letting us have a few, they refused to part with any more, even for a good price, assigning as a reason, that they belonged to a man who was absent. They either prize that kind of food very highly, or are scrupulous about using the property of an absent countryman.

A heavy gale continued all day, and raised a very high surf on the beach. As the weather was extremely cold, we required a considerable quantity of firewood, and converted two of the Eskimo scaffolds to that use; but I informed the owners that we would pay for it before we went, to which arrangement they gave a ready assent. At high-water, about a quarter before seven o'clock, A.M., the sea rose so much that the shoal off the point was covered, and the surf began to reach our boats, so that we had to shift them further in. The wind blowing on the shore had increased the rise of water considerably beyond an ordinary tide. It was low-water again a few minutes before one. In the afternoon the gale began to moderate, and in the evening three men came to the village, two of them being of the party who had visited the boats three days before. They told us that their women were coming in the umiaks on the following day.

We had now a pretty numerous body around us, and all were perfectly orderly except the man that we first saw, who seemed to be the chief, or perhaps the *shaman* or conjuror, of the community. His features were forbidding, but the younger men had intelligent countenances, modest and cheerful manners, and were neither forward nor troublesome. They were of moderate stature.

August 7th. — The wind continuing to abate throughout the day, and the sea to subside, we prepared to resume our voyage as soon as the evening tide had flowed sufficiently to admit of the boats being launched over the sand-flats. In the afternoon we saw the women and children approaching slowly in the umiaks. They stopped about a mile off, and, notwithstanding the signals and shouts of their husbands, hung back. The men called out *umiët kai-it,* "Boats, come here," with a peculiar elevated intonation, which could be heard at a great distance, similar to that practised by the inhabitants of the Swiss Alps. We did not await their arrival, but, having found water enough about half a mile to the westward of the point, pushed out to sea; and partly under sail, partly by rowing, coasted during the night a very low shore, varied by a few higher islands, which the very flat sands surrounding them prevented us from approaching.

At eight in the morning of the 8th of August we landed on one of a cluster of clayey and sandy islands, near Cape Brown; and while we were preparing breakfast were visited by four Eskimos, who told us that they had just killed several reindeer on a neighbouring island. On desiring to have some of the venison in exchange for a knife they went away in their kaiyaks, and, before we had finished breakfast, returned again, bringing with them only the toughest and most inferior parts of the animals. I paid them what was promised, and made them some presents besides, but reprimanded them for their want of hospitality, and told them that I meant to have given them an axe had they been more liberal. These people, like the other parties we had previously communicated with, declared that no large ships nor boats had been seen on their coasts, and that we were the first white men they had ever beheld. I could not discover that any remembrance of my visit to their shores, twenty-three years previously, existed among any of the parties I saw on the present voyage, though I never failed to question them closely on the subject.

After breakfast we crossed Russell Inlet, and as we passed Cape Brown several Eskimos put off from the shore, and three of them overtook us though we were going with a stiff breeze at the utmost speed of our boats, or about six geographical miles an hour. To those that came up we made presents, and put the usual questions.

In the evening we anchored the boats under the westernmost of two islands lying immediately off Cape Dalhousie, and, having landed, pitched our tents on the beach. The islands here, and the Cape itself, consist of loam or sand, and present steep cliffs towards the sea forty or fifty feet high. The surface is level, except where ravines, occasioned by the melting of the snow in the beginning of summer, intersect it; and all the islands are so surrounded by sandy flats that a boat cannot come near the beach. On the present occasion we anchored three or four hundred yards from the shore.

The island on which we encamped is a breeding-place of the *Xema sabinii*, the handsomest of all the gulls. Many of the parents were flying about accompanied by their spotted young, also on the wing. This is the most westerly ascertained breeding station of the species, which has been found at Spitzbergen, Greenland, and Melville Peninsula. Mr. Rae shot some fine male specimens, whose plumage and dimensions agreed exactly with the description in the *Fauna Boreali-Americana*. The eggs are deposited in hollows of the short and scanty mossy turf which clothes the ground.

August 9th. — Through the carelessness of the

night-watch the boats were suffered to ground on the ebbing of the tide, and it was not till eight o'clock that the water had flowed sufficiently to permit of our embarking.   We then stood across Liverpool Bay with a very light breeze, and about two o'clock had sight of the eastern coast near Nicholson Island, the western shore being visible at that time, but soon afterwards sinking below the horizon.   At half-past nine, P. M., we reached the eastern shore, and encamped under the frozen cliffs of Cape Maitland.   This cape is an island, and on the former voyage I passed through the channel which divides it from the main.   Its surface is nearly level, the soil is loam or clay, and the cliffs which bound it are about eighty feet high, and, being worn at the base by the action of the waves, overhang the narrow beach by eight or ten feet.   Landslips are frequent, and occasion a frozen surface to be constantly exposed to view.   The island does not differ from the neighbouring lands in its subsoil being frozen : since permanent ground ice is found everywhere at eighteen or twenty inches beneath the surface.   All access from below would have been cut off by the overhanging cliff, were it not for deep gullies which here and there afford a steep path to the top.   Vegetation was very scanty, and throughout this voyage I observed that flowering plants were more scarce, and the herbage generally thinner, than in 1826. The bulk of the cliff is composed of a black clay or loam, which is disposed in undulated beds, and, in some places, the section exhibits a spherical mass eight or ten yards in diameter, with concentric layers, like the coats of an onion. A few pebbles occur in the loam, and the beach is formed of sand and small pebbles washed from the cliffs, and consisting mostly of trap mixed with quartz, and a little white sandstone and limestone.

In 1826 I observed many slabs of red sandstone in the channel behind the island, but there were none at our present encampment. I caused a pit to be dug at the top of the cliff, and found that the thaw had penetrated sixteen inches. A thermometer laid on the bottom of the pit indicated 33°F., the temperature in the shade being at the time 42°F. High-water occurred at 1 A.M., and the ebb flowed to the northward along the island.

The distance we ran from our encampment at Cape Dalhousie to Point Maitland, measured by Massey's log, was thirty-five geographical miles, or thirty-one direct (excluding the angle we made), which agrees exactly with Lieutenant Kendall's map of 1826. Mr. Rae killed some rock ptarmigan in the night. The old males were moulting at this time.

August 10*th*.——Having breakfasted before embarking, we left our sleeping-place at 6 A.M. Soon afterwards we saw two Eskimo tents on the extreme point of the island; but as the inmates did not show themselves, and we did not wish to be delayed, we proceeded onwards without disturbing them, and crossed the mouth of Harrowby Bay. At 4 P.M. Baillie's Islands came in sight, and we held on our course between them and the main. Some Eskimos coming off here, we learned from them that a large river falls into the bottom of Liverpool Bay; and we had previously received the same information from the party which we saw on the 8th. Eskimos inhabit the banks of this river, but the families residing at Cape Bathurst do not go so far. The river can be no other than the " Begh'ula tessè " of the Hare Indians, who frequent Fort Good Hope. These Indians say that it is a large river, abounding in the fish from which it is named (" toothless fish," *Salmo Mackenzii*); that it rises near Smith's Bay of Great Bear Lake, and is eight or ten days' journey to the eastward of Fort Good Hope overland (one hundred miles or thereabout). They also said, that some years ago they fell in with a party of Eskimos who were hunting on its banks, and, a quarrel ensuing, several of the latter were killed.

We could not find a convenient landing-place

either on Baillie's Islands or on the main shore, owing to the flatness of the coasts, and were compelled to anchor the boats nearly a mile from the beach. The men waded ashore to collect driftwood and cook supper, after which we all embarked to sleep in the boats. We had scarcely completed our arrangements for passing the night, when we became aware of a fleet of kaiyaks with three umiaks coming down upon us in a crescentic line, looming formidably in the faint twilight. As I did not wish the men's rest to be interrupted by visiters, a ball was fired across their path to arrest their progress, on which they assembled in a group evidently in consultation. Albert now hailed them by my direction, and said that we were going to sleep, and that the sentinels would fire on any one who came near the boats by night, but that we should be glad to trade with them in the morning. When they fully understood our wishes, after a little further parley, they retired and did not afterwards trouble us.

At 2 o'clock A. M., on the 11th, we weighed anchor, and nine Eskimos, with their kaiyaks, and three umiaks, containing the women, having come off, they led us by a sufficiently deep channel to the westward of a dry bank or sandy island, over a bar on which we found from four to five fathoms of water. On crossing this we passed suddenly from muddy water into a green sea, in which we had no bottom with the hand-lead. In 1826 we sought a way out to the eastward of the island, and, the bar there being muddy and shallow, found no little difficulty in forcing the boats through. The northern channel would form a good ship harbour.

From these people we learnt that during their summer of two moons they see no ice whatever, that they were now assembling to hunt whales, and would go out to sea to-morrow for that purpose. The black whale, their present object, they call *ai-ë-werk*, and the white whale, which also frequents this coast, *keilaloo-ak*. In some summers they kill two black whales, very rarely three, and sometimes they are altogether unsuccessful. In the course of conversation, we were told that the several families have hunting grounds near their winter houses, on which the others do not trespass; and the proprietors of several points of land in sight were named to me. They knew but little of the country beyond their own vicinity; and one of them having told me that Cape Bathurst was an island, I affirmed that it was not, on which, with an air of surprise, he exclaimed, " Are not all lands islands ?" None of them could remember my former visit, though I had communicated with a party of their countrymen only a few miles from their present residence. We told them that we were looking for ships and men of our nation whom we expected to meet; and they said they would be glad of the visits of white men, and would treat them hospitably. In exchange for some fish, seal and whale-skin leather, and a few other things, we supplied them well with knives, files, hatchets, and beads. Part of the number who wished to come to the boats the preceding evening had, on our declining the interview, gone to their winter houses on the western shores of Baillie's Islands; and those who accompanied us from our anchorage in the morning landed on the extremity of Cape Bathurst, where their winter houses stood.

It was part of my instructions to bury some pemican at this cape, and to erect a signal-post; but the presence of the natives hindered me from doing so. As soon, however, as we had gone far enough to be, as we supposed, beyond their view, we put on shore, and having dug a hole on the top of the cliff, deposited therein a case of pemican, with a memorandum explaining the objects of the expedition. Every precaution was used to replace the turf, so as not to betray that it had been moved; some drift-timber was piled upon it and set on fire, and a pole, painted red and white, planted at the distance of ten feet. As all the drift-timber at this place had been gathered into a heap by the Eskimos, and the pole was part of it, we hung up some articles of value to them by way of payment, in the hope that it would cause them to respect the signal-post. In the meantime our crews were preparing breakfast, and we had just finished this meal, when we saw some Eskimos from the cape running towards us. They had evidently been watching us, and came in the expectation of receiving some additional articles; nor were they disappointed. The soil here was thawed to the depth of fourteen inches. The deposit was made about five miles from the extreme point of Cape Bathurst.

Many black whales and two white ones were seen this morning. The eider ducks had now assembled in immense flocks, and with the brent geese were migrating to the westward. Both these water-fowl follow the coast-line in their migrations on the Pacific as well as the Atlantic sides of the Continent. The eiders are only accidental visitors in the interior, and the brents are not seen inland to the eastward of Peel's River; but Mr. Murray informs me that in their northerly flight they follow the valley of the Yukon, thus cutting across the projecting angle of Russian America.

The surface of the country in the vicinity of Cape Bathurst is level or gently undulated, and the sea cliffs are in many places nearly precipitous, and about one hundred and fifty feet high. The strata, where exposed, are sand and clay, and I believe that this promontory, from its northern point to the bottom of Franklin Bay, is the termination of the sandy and loamy deposit and bituminous shale which throughout the whole length of the Mackenzie rests on the sandstone and limestone beds so frequently noticed in the preceding pages, and fragments of which may be traced among the alluvial islands in the estuary of the Mackenzie, and in Liverpool Bay. A line drawn from *Clowt sang eesa*, or Scented Grass Hill, of Great Bear Lake, to the north-north-west, would form a tangent to the eastern coast-line of Cape Bathurst, and most probably mark the limit of the formation on that side. If so, the River Beghula, which enters Liverpool Bay, will flow through a country similar to that forming the banks of the Mackenzie, and being consequently well wooded, will abound in animals.

As we proceeded to the south-east from Cape Bathurst along the shore, the crest of the high bank rose to about two hundred and fifty feet, and beds of bituminous shale, similar to those on the Mackenzie, are exposed in many places. At Point Trail, in latitude 70° 19′ N., the bituminous shale was observed to be on fire in 1826, and the bank had crumbled down from the destruction of the beds. Selenite, alum in powder, and the wax-coloured variety of that salt named " Rock butter," with sulphur, were among the products of the decomposition of the shale which I then collected; and the clays which had been exposed to the heat were baked and vitrified, so that the spot resembled an old brick-field. The sand covering the shale here is coherent enough to be a friable sandstone; and many concretions of clay iron-stone exist in the shale beds, exactly similar to those which are imbedded in the shale of Scented Grass Hill. Wilmot Horton River flows out by a narrow gorge from a flat valley, and the high banks, rising in ridges above the valley, flank it some way inland, as we had noticed them doing on the tributary streams that join the Mackenzie.

Near Point Fitton the cliff is two hundred feet high, and contains layers of rock-butter two inches thick, with many crystals of selenite adhering to the surface of the slates. The cliff is capped by a marly gravel, two yards thick, containing pebbles of granite, quartz, Lydian stone, and compact limestone. To the southward of Wilmot Horton River, portions of the ruined bank continue to emit smoke.

Cape Bathurst has been recently invested with more interest since it is the point of the main shore from whence Commander Pullen was directed by the Admiralty to take his departure in the summer of 1850, in his adventurous attempt to reach Melville Island. By the last accounts from Mackenzie's River, we learn that this enterprising officer received his instructions by express, on the 25th of June, being then in Slave River, on his way to York Factory. He immediately turned back, having been supplied with 4,500 lbs. of jerked venison and pemican by Mr. Rae, which he embarked in one of the Plover's boats, and in a barge of the Hudson's Bay Company, being the only available craft. The barge is well adapted for river navigation, but from its flatness unfitted for a sea-voyage, though it may be in some respects improved by the addition of a false keel, which Commander Pullen would probably give it before he descended to the sea. Its weight will render it much less manageable among ice than a lighter boat. No intelligence of this party has reached England since the above date, but we may expect to hear of his proceedings in May or June, 1851, before this volume has passed through the press.

* This anticipation has been realised, as has been mentioned in p. 216. Commander Pullen found the sea covered with unbroken ice all the way from the Mackenzie to Cape Bathurst, a small channel only existing in shore, through which he advanced to the vicinity of the cape. Failing in finding a passage out to sea, to the north of Baillie's Islands, he remained within them, until the advance of winter compelled him to return to the Mackenzie.

CHAP. IX.

VOYAGE CONTINUED ALONG THE COAST. — FRANKLIN BAY. — MELVILLE HILLS. — POINT STIVENS. — SELLWOOD BAY. — CAPE PARRY. — COCKED-HAT POINT. — CACHE OF PEMICAN. — ICE PACKS. — ARCHWAY. — BURROW'S ISLANDS. — DARNLEY BAY. — CLAPPERTON ISLAND. — CAPE LYON. — POINT PEARCE. — POINT KEATS. — POINT DEAS THOMPSON. — SILURIAN STRATA. — ROSCOE RIVER. — POINT DE WITT CLINTON. — FURROWED CLIFFS. — MELVILLE RANGE. — POINT TINNEY. — BUCHANAN RIVER. — DRIFT ICE. — CROKER'S RIVER. — POINT CLIFTON. — INMAN'S RIVER. — POINT WISE. — HOPPNER RIVER. — WOLLASTON LAND. — CAPE YOUNG. — STAPYLTON BAY. — CAPE HOPE. — CAPE BEXLEY. — ICE FLOES. — POINT COCKBURN. — A STORM. — CHANTRY ISLAND. — SALMON. — LAMBERT ISLAND. — LEAVE A BOAT. — CAPE KRUSENSTERN. — DETAINED BY ICE. — BASIL HALL'S BAY. — CAPE HEARNE. — PECULIAR SEVERITY OF THE SEASON. — CONJECTURES RESPECTING THE DISCOVERY SHIPS. — RESOURCES OF A PARTY ENCLOSED BY ICE AMONG THE ARCTIC ISLANDS. — GENERAL REFLECTIONS.

August 11th, 1848.—WE sailed along the coast all day with a light breeze, and in the afternoon eleven Eskimos came off from the shore and sold us some deers' meat. A woman of the party ran for two miles along the beach in the hope of receiving a present, and, when quite exhausted with her exertions, stripped off her boots to barter with us. One of the men in the kaiyaks brought them off, but, as they were too small for any of our crew, we returned them with a present of more than their

VOL. I.   T

value. These men gave us no additional information, but expressed pleasure when told that they might expect to meet other parties of white men. A scull of eleven white whales were seen in the evening.

We continued under sail all night, and at three in the afternoon of the 12th, landed in a very shallow bay, to the southward of Point Stivens, to cook a meal which served for both breakfast and dinner. Mr. Rae went in pursuit of some rein-deer which were seen from the boats, but, owing to the extreme flatness of the land, which afforded no cover, he was unable to approach them.

The high banks of Cape Bathurst are continued to the bottom of Franklin Bay, where they recede a little from the coast, and are lost in an even-backed ridge, apparently not exceeding four or five hundred feet in height. These hills are named the Melville Range, and cross the neck of the peninsula of Cape Parry, appearing again behind Darnley Bay. The peninsula is so flat near its isthmus, and so much intersected by water, that I am still in doubt whether it may not be actually a collection of islands. But if this is the case, the channels which separate the islands are intricate and shallow. To the south of Point Stivens the soil was wholly mud, apparently alluvial; to the northward beds of limestone crop out. In the evening we encamped on Point Stivens, which is a long narrow gravel beach, composed mostly of pieces of limestone, some of which contain corals. Sea-weed is very scarce throughout the whole of the arctic coast, but we saw on the beach here some rejected masses of decayed *Laminaria*, probably *saccharina*; also a stunted white spruce, lying on the beach, still retaining its bark and leaves. Mr. Rae shot a fine trumpeter swan, on which we supped. The only water we could find here for cooking was swampy, and full of very active insects shaped like tadpoles, which were just visible to the naked eye (*Apus*; *Lepidurus*, probably *Lynceus*).

The crimson and lake tints of the sky, when the sun set this evening, were most splendid, and such as I have never seen surpassed in any climate.

On August the 13th, we embarked at 3 A. M., and at half-past ten landed in Sellwood Bay on some horizontal beds of limestone, which are the first rocks *in situ* that become visible, in tracing this peninsula from the south. No organic remains were detected in the stone. Many very large slabs, moved but a short way from their parent beds, were piled upon each other within reach of a high surf, and among them lay great boulders of greenstone-porphyry and hornblende rock. To the north of the bay, there are high cliffs of limestone, and also a detached perforated rock, which employed

T 2

Lieutenant Kendall's pencil on my former voyage. Many white-winged silvery gulls were breeding on the various shelves of its cliffs, and their still un-feathered young were running about, alarmed by the clamour of the parent birds.

In the evening we anchored in a snug boat-harbour, within the westernmost of the two points which terminate Cape Parry. The part of the Cape which will be first visible on approaching from sea, is a hill about five hundred feet high, which far overtops all the neighbouring eminences. From it a comparatively low peninsular point stretched west-north-west about half a mile, being connected to the main by a gravel bank, and ter-minated on its sea-face by a limestone cliff, which in some points of view resembles a cocked-hat. An indented bay, about three miles across, sepa-rates this point from another more to the east, which extends fully as far north. Booth's Islands, five in number, form a range nine miles long, whose extremities bear from the hill north-west and south-west respectively. The channels be-tween the islands vary in width from one to three miles. On the east side of the hill, cliffs of lime-stone, washed by the waves, have been scooped into caves and arches, which, without much aid from the imagination, recalled many fine archi-tectural forms. A boulder of chert, lying on the shore, measured five feet in length, by four in breadth, and exhibited some curious veining.

In approaching our anchorage we shot two seals, and one of them, being thrown on the strand, attracted, late in the evening, the attention of a grey fox, which was prowling along the beach. As soon as the animal saw the carcase, it halted, and, after a momentary survey, leaped lightly behind a large log of drift timber, from whence it peeped out at it from time to time. While I was watching the fox's reconnoitring tactics with some interest, and speculating upon the time it would devote to the survey before it ventured to approach the carrion, a noise made by the sentinel, who was seated by the fire, scared the wary object of my study, and it fled swiftly up the hill.

August 14*th.*——This morning we deposited a letter, with a case of pemican, on the verge of the cliff of Cocked-hat Point, covered it with fragments of limestone, and erected in front of it a pile of stones marked with red paint. The beds of this cliff are horizontal, and they are not only interleaved with regular layers of chert, but also contain nodules of the same material. On rounding the eastern point of Cape Parry, we saw packs of drift-ice for the first time since we commenced our sea voyage. Eight or ten miles to the southward of the Cape we passed an isolated rock, perforated by an archway, standing in front of some bold limestone cliffs; soon after noon we were opposite to a point distinguished by a remarkable rounded hummock; and in the evening we encamped on one of Burrow's Islands. This island is composed of cherty limestone, which in decaying acquires a honeycombed surface as hard as a flint or file. Rain fell in the night for the first time for many days.

A fair wind having sprung up in the night, we embarked an hour before midnight, and on the morning of the 15th stood down the bay, with light breezes and hazy weather; landing at 10 A. M. on Clapperton Island to prepare breakfast. The coast in this quarter is similar in character to that on the opposite side of the peninsula in the same parallel, being low and not easily approachable on account of the extensive sandy banks which lie off it. Clapperton Island, itself, is gravelly. From it, we saw land round the bottom of the bay, with some intervals apparently inlets.

On re-embarking, we steered directly for Cape Lyon, distant about ten miles; but a low thick fog coming on, we got involved in a stream of drift-ice, on which the course was altered a little, so as to fetch within the cape. We made our way through the ice without damage to the boats, and in the afternoon found ourselves about four miles from the pitch of the cape, close in-shore, under some very high mural precipices of a bluish-grey, slaty rock, on which a thick mass of columnar basalt is imposed. The cliffs are about two hundred feet high, and stand out in succession, forming the salient angles of several shallow indentations of the coast-line. A talus of unmelted snow-drift lay under most of them, which, being undermined by the action of the waves, would be detached on the first heavy fall of rain and become icebergs. Towards the bottom of Darnley Bay, the coast-line declines greatly in altitude, but the heights of the Melville Range are dimly seen in the distance. We anchored for the night in the north-west angle of the cape, between two projecting points of basalt. The land here rises betwixt three and four hundred feet above the sea.

On the 16th we continued our voyage to the eastward, and, on landing to prepare breakfast at Point Pearce, ascertained that high-water took place there at eight o'clock. The trap ranges run in this quarter in a south-west and north-east direction, and produce small bays; their precipitous faces are turned towards the west-south-west. Judging from the view we had of them from the boats in passing, the cliffs were mostly composed of greenstone-slate. At Point Pearce, the shore is formed of flesh-coloured limestone, whose beds crop out in successive cliffs like stairs, and attain a height of two hundred and fifty feet a short way from the beach. At a small point lying between the one just named, and Point Keats, there are magnificent columns of basalt with a pillar of the same material rising out of the water immediately in front of them. The bay to the south of these columns is lined with cliffs of flesh-coloured limestone; thin layers of sandstone crop out further to the eastward, and covering them there is an overflow of dark leek-green basalt or greenstone.

A strong head-wind having sprung up and occasioning much fatigue to the rowers, while our progress was small, we put in at Point Keats, and encamped. Mr. Rae and Albert went to hunt rein-deer, and I took a short walk inland. I soon came to extensive beds of flesh-coloured sandstone, forming the bounding walls of a deep narrow ravine, through which flowed a small shallow river varying in width from ten to thirty yards, but whose channel bore evidence of a considerable body of water passing through it in the spring floods. The ridges of sandstone seem to have a direction from west-north-west to east-south-east, or nearly parallel to the general coast-line here, and are much fissured, the principal fissures being in the line of the strike. The cliffs for the most part face westerly, and the west wall of the ravine is lower than the eastern one. The sandstone is fine-grained, hard, and durable. Some beds are very white, others flesh-coloured, and interleaved with them are beds of chert or quartz rocks. The dome-shaped summits of the Melville Range are visible in the distance.

The fragments of sandstone cover many miles of surface, and the limestone, which splits off in thin layers by the action of the frost, becomes by the action of the weather as rough as a file, and soon wears out a pair of shoes. There are some beautiful excavations in the cliffs that are exposed to the waves, and a fine gothic shrine, with a canopy and mouldings supported on slender pillars, attracted our special notice. An isolated column, which stands before it, had been selected as a breeding-place by two ivory gulls, who were very clamorous when any one approached their nest. The young had ash-grey backs, and were nearly fledged. A very clear sunset enabled me to obtain an extensive view to seaward, and I am convinced that no land much above the level of the water lies within forty miles to the northward of this point.

A thick, wet fog, accompanied by a strong head-wind, detained us at our encampment till after breakfast on the 17th. During our enforced stay, Mr. Rae killed a roe rein-deer in excellent condition, and we procured also some waveys (*Anser hyper-*

boreus).   These geese and the northern divers (*Colymbus borealis*) were at this time migrating to the south-east, or in the opposite direction with respect to the line of coast that we had observed the eider ducks and brent geese proceeding a few days previously near Cape Bathurst.

About a mile and a half to the westward of Point Deas Thomson, the projecting point of a deep cove is perforated, forming a natural bridge, and not far from it another projection exhibits a less striking opening.

Torso Rock.

A detached column of limestone, enclosing masses and layers of chert or quartz rock, is also cut through, forming a pointed arch.   The whole coast is composed of limestone, forming high cliffs at intervals. The quartz rock beds acquire occasionally a pistachio-green colour, as if from the presence of epidote. A similar stone occurs at Pigeon River on the north shore of Lake Superior; and the limestones and sandstones of the latter district, with their associated trap rocks, as at Thunder Mountain, correspond in most respects with those between Cape Parry and the Coppermine River; and consequently, if we can rely on lithological characters, they may be considered as the oldest members of the silurian series, or as the rocks on which that series is deposited, to which epoch the Lake Superior formation has been assigned. If we had been able to trace up the limestone spurs of the Rocky Mountains which traverse the Mackenzie, they would most probably have been found running up to and connected with the limestone of this coast.

At six in the afternoon the wind veering round to the south-west, we embarked again, and continued under sail till three in the morning of the 18th, when the wind failing us we dropped anchor in the mouth of Roscoe River. We resumed our voyage at 8 A.M., after preparing an early breakfast.

A little to the westward of Point de Witt Clinton, a range of basalt and limestone cliffs extends for a mile along the beach, under which there lay a talus of unmelted drift-snow.   In 1826, though we passed along this coast nearly a month earlier in the season, we observed much fewer of this kind of memorial of the preceding winter, and that year, the flowering plants were more plentiful, and the vegetation generally more luxuriant.   The deterioration of the climate after rounding Cape Parry became daily more and more evident to us, though we had decreased our latitude above a degree since leaving Cape Bathurst. The presence of a large body of warmer water, carried into the Arctic Sea by the Mackenzie, may have some influence in ameliorating the temperature within the limits of that wide estuary; but I am inclined to think that Wollaston and Banks's Lands, by detaining much drift-ice in the channels which separate them from one another and from the main shore, have a still more powerful effect in lowering the summer temperature.

At Point de Witt Clinton the cliffs are formed of flesh-coloured beds of limestone interleaved with bluish-grey beds, and containing fibrous and compact gypsum in veins.   These cliffs are forty or fifty feet high, and are covered to a considerable depth with diluvial loam, containing fragments of sandstone, limestone, and trap rocks, some of them rolled, others angular.   The surface of the loam is undulated; and, about a quarter of a mile from the beach, cliffs of basalt protrude, at the height of two hundred and fifty feet above the water. A short way to the westward also of the point, cliffs of basalt rise from the beach. This stone breaks up here into cubical blocks, many of which are piled up at the foot of the cliff. These fragments and the basaltic shelves at the base of the cliff, are sculptured by fine acute furrows, and polished by the action of ice and gravel, the scratches being generally perpendicular to the line of coast, but occasionally crossing each other. The Melville Range is about five miles distant from this part of the shore, and presents many mural precipices and ravines on its acclivity. The highest points did not appear to rise more than seven or eight hundred feet above the sea. An undulated grassy country intervenes between the range and the shore.

In the evening we encamped on a point situated in latitude 69° 30′ N., to the westward of Point Tinney. The sea-bank shelves down from the general level of the country here one hundred and fifty or two hundred feet, and, being cut by ravines, shows conical eminences when seen from a boat. The diluvium is at least forty feet thick.

August 19*th*.——This morning we crossed the mouth of Buchanan River, which is a very small stream in this month, but the channel which it fills in time of flood is one hundred yards wide. I have mentioned, in the preceding narrative, that when we visited Fort Good Hope, no rain had fallen there this season, and a few short showers only occurred after we came to the coast. The banks of this stream gave further evidence of the dryness of the summer, in the clayey soil being cracked every where into round flat cakes, on which the foot-prints of geese, which had walked over them while yet muddy from the melting snow, were sharply impressed. Were such a surface to be covered by drift-sand, the foot-prints might be preserved as in the ancient sandstones. In the present instance the winter frosts set in without any heavy rains having fallen to obliterate the traces, which would consequently remain hard until the following spring.

Mr. Rae brought in two fine rein-deer, and several seals also were killed; but none of the men relishing the dark flesh of the seals, while they had abundance of excellent venison, I gave directions that no more should be shot. A meridional altitude was obtained in lat. 69° 19½′ N.

To-day we passed through much drift-ice by very devious channels, and not without risk of the boats being crushed, but fortunately without damage. Croker's River issues from a triangular, level valley, three or four miles wide at the beach, and extending about five miles backwards. Over this the stream spreads when flooded, but when low, filters out by narrow channels, barred across by sand-banks. The valley is bounded on the east and west by elevated banks of sand, diluvial loam and boulders, which meet at the Melville Range. The valley at this time presented a singular scene of desolation; for though the summer was now far advanced, its flat bottom was entirely covered with large floes of ice, which had been probably driven over the sand-bar from the sea by northerly winds.

A brood of long-tailed ducks (*Harelda glacialis*) were seen swimming in one of the streams with the mother bird in the van. Her wariness did not prevent us from laying her flock under contribution for our evening meal.

August 20*th*.—This day also our voyage was performed among crowded floes of ice, and was consequently slow. When we landed to prepare breakfast, Mr. Rae killed a fine buck rein-deer. In this quarter, a skilful hunter, like Mr. Rae, could supply the whole party with venison without any loss of time. A meridional observation was obtained in lat. 69° 9′ N. between Point Clifton and Inman's River, and about two miles from the latter; the variation of the compass, by the sun's bearing at noon, being 61½° E., and Point Clifton bearing north 26° west, distant a mile and a quarter.

A little to the eastward of Point Clifton, there are cliffs of limestone, from whence to Inman's River the beach is alluvial and shingly.   The river flows between high gravel banks and alluvial cliffs ; and to the eastward of it, the limestone rises in successive terraces to the height of four hundred feet above the sea.   The eminences which this formation produces are long and round-backed, and it abounds in narrow deep ravines or fissures.

At Point Wise, the cliffs are composed of crumbling earthy limestone, containing chert in layers and nodules.   From this point, at sunset, Wollaston Land was distinctly seen at the distance apparently of thirty miles.

On the morning of the 21st, we passed two ranges of high limestone cliffs, at the second of which, lying to the eastward of Hoppner River, we put ashore to prepare breakfast.   The ice under this cliff was loaded with many tons of gravel.   Wollaston Land, as seen from hence, appeared to have its summits and ravines covered with snow, but the channel being filled with ice, the ice-blink rendered the true form and condition of distant land very uncertain.

One of our boats having been injured by the ice and rendered leaky, we put ashore early at Cape Young to repair the damage, which was effected in the course of the evening.

August 22*nd.*—On embarking this morning in rather thick weather we struck across Stapylton Bay, for three hours, and then, getting sight of Cape Hope, bearing east-north-east, hauled up for it.   The sky was dark and louring, with occasional thick haze and heavy showers, and a water-spout, seen in shore, gave intimation of an approaching storm.   Ice floes lying close off Cape Hope caused us no little trouble, the passages among them being very intricate, and the perpendicular walls of the masses being too high to allow of our landing, or seeing over them.   In the afternoon we passed Cape Bexley, running before a stiff breeze, and at 5 P. M., a storm coming suddenly on, we were compelled to reduce our canvass to the goose-wing of the mainsail, under which we scudded for an hour, and then entering among large masses of ice, about two miles from Point Cockburn, found shelter under some pieces that had grounded.   The shore was too flat to admit of our bringing the boats near enough to encamp; the ice-cold sea water chilled the men as they waded to and fro; there was no drift-timber on the beach; and we passed a cold and cheerless night in the boats, the wind being too strong to admit of our raising any kind of shelter. I afterwards learned that this storm began at Fort Simpson at 6 A. M. on the 23rd, or, making allowance for the difference of longitude, about thirteen hours and a half later. It commenced on the Mackenzie by the wind changing from north-east to north-west, and the sky did not clear up till nine in the morning of the 24th. At the same date an earthquake occurred in the West India Islands, which did much damage.

During the night much ice drifted past, and in the morning of the 23rd the sea as far as our view extended was one dense close pack. By 10 A. M. the wind had moderated considerably, and the rising tide having floated some of the stranded pieces of ice, we were enabled to advance slowly along the shore by moving them aside. In this way every small indentation of the coast-line required to be rounded, and as these were numerous the direct distance made good was small. We encamped, on the tide falling again, at 2 P.M., on a gravel point lying about ten miles to the westward of Chantry Island. Snow, which fell in the night, did not wholly melt this day, and the distant rising grounds were white. The weather continued very cold; drift-wood proved to be exceedingly scanty; and in the night we had high winds and much sleet. The coast-line is more deeply indented in this quarter than the chart * indicates, as chains of low sandy islands which lie across the entrances of the bays hid them from our view on the former

* In Franklin's second Overland Journey.

voyage.   The country is flat and strewed with fragments of limestone.

No lanes of open water could be discerned on the 24th from any of the eminences near the coast. By handing the boats over the flats, where the water was too shallow for heavy ice, we were enabled to round six small bays.   In the course of this labour many salmon were seen, a few were killed by the men with their poles, and some were found on the ice, having been left by seals, which were scared away by our approach.   The shape of this salmon is much like that of the common sea-trout of England, but its scales are rather smaller.   Its flesh is red, and its flavour excellent.   A medium-sized fish measured 29 inches in length, and 16 in girth.   We encamped at five in the evening a little to the eastward of Chantry Island, having travelled about twenty-four miles round the bays, but gained only eight in direct distance.   At this place, beds of compact white limestone crop out on the beach, and the surface of the country is thickly strewed with boulders of bright-red sandstone, some of them very large.   Many boulders of basalt and other trap rocks also occur.

August 25th. — A strong west-north-west wind blowing during the night, cleared away much of the ice that pressed immediately on the beach, though all remained white to seaward. We were enabled thereby to run for three hours before the wind, but then came to a bay, through which there was no passage, large floes resting on the rocks of the beach, and no lanes existing outside. A meridional observation gave the latitude of this place 68° 36′ N. Lambert Island lies some miles distant in the offing. The surrounding country, as far as my examination extended, consists of limestone, but many sandstone boulders of various colours, lying on the surface, point out that stone as existing *in situ* in some locality not far distant.

August 26th.——A frosty night covered the sea and ponds with young ice, and glued all the floes immoveably together, so that the rise of the tide was no longer serviceable to us. We carried the cargo and launched the boats across a point of land for half a mile in the morning, and spent the rest of the day in the various operations of cutting through tongues of ice, dragging the boats over the floes, where they were smooth enough, moving large stones that lay in the way, and resorting to every expedient we could devise to gain a little advance. Two more portages were made in the afternoon over rugged paths, and we travelled in all about five miles in a day of very severe labour. A heavy snow-storm converted the surface of the pools of sea water into a thick paste, the water being already cooled down to the freezing point.

By a repetition of the same operations, which occupied us during the previous day, we advanced on the 27th about three miles and a half. After Mr. Rae had attentively examined the sea from a high cliff without perceiving any slackness in the ice or motion during the flood or ebb tide, I determined on lightening the labours of the men, by leaving one boat and her cargo on a rocky point, which bears north 28° west from Cape Krusenstern, distant twelve miles. Our encampment in the evening was on a flat terrace of slaty limestone under a high cliff of the same rock. The limestone reposes on beds of chert or quartz rock in thin layers, which in some places are detached in large slates. Here we deposited, on a flat shelf of the rock, several cases of pemican, an arm-chest, and some other things that encumbered the boats, and rendered them less fit for launching over the ice.

During the night, a fresh wind from the east-south-east brought much snow, which added to the pasty condition of the surface of the water, and produced a layer of semi-fluid matter that completely deadened a boat's way under oars.

Three hours were consumed on the morning of the 28th in bringing the boats about a hundred yards, the cold weather almost paralysing the men's powers of exertion.

When the tide flowed in the afternoon, a portage of a thousand yards was made, and the boats being afterwards dragged across some smooth floes of ice, we gained a pool of open water, and pulled to the bottom of Pasley's Cove, where we encamped. Mount Barrow is a conspicuous object from the bottom of the cove, as it rises abruptly from the flat limestone strata.

August 29*th.*——During the night and this morning the same keen frosty east winds continued, with snow showers. The tide fell so much that the boats were left aground in the morning, and we were unable to proceed until 7 A.M. In three hours we came within about a mile and a half of the pitch of Cape Krusenstern, when further progress being barred by ice heaped against the cliffs, we put ashore and drew up the boats on the beach. In the flat limestone beds, of which the country here is formed, I observed a curious variety of structure which I saw no where else, and which I cannot satisfactorily account for. It was a diminutive ridge like the roof of a house, formed in this manner ∧, of the upper slaty layer of the limestone, its height or the breadth of its sides being about a foot only, yet its length was half a mile. It seemed to be connected with a fissure. Though the layers of the limestone are most extensively detached by the freezing of the moisture, which insinuates itself between them, such a process could not produce any thing so regular as this small anticlinal ridge. If beds already fissured were, in subsiding, to be pressed more closely together, the edges of the fissure might perhaps assume such a form. Elsewhere, in the same formation, straight furrows, as if drawn by a plow, are common, and evidently proceed from the small fragments which cover the ground, filling a crack, nearly to the brim. Judging from the whole surface here being covered with thin pieces of limestone to the exclusion of soil, I should infer that the frost splits off the layers and breaks them up more effectively than any agent to which rocks are exposed in warmer climates, and that the scantiness of the soil is owing to the shortness of the season of growth of the lichens and other plants, which have the power of decomposing the surface of the stones and so producing a little mould. The frost breaks up the stone before the lichens have time to establish themselves.

The limestone which forms the cliffs of Cape Krusenstern, and the other cliffs on the coast between it and Cape Kendall, contains many thin slaty beds of chert or quartz rock, either bluish-white, or coloured reddish-brown by oxide of iron.

We remained all the 30th in an encampment, watching the ice outside, or making excursions across the cape to examine the sea in various directions. Some small lanes of water were visible, and the ice was moved to and fro by the flood and ebb, but no channel was discovered by which we could hope to make any progress towards the Coppermine River. The wind continued in the east-north-east quarter, and the weather was very chilling. We employed the men in erecting a column of stone near the tents. It was on this cape that Mr. Rae spent a month of the following summer in anxiously watching for an opening in the ice, by which he might cross to Douglass Island and Wollaston Land. The true position of Douglass Island is ten miles from Cape Krusenstern.

At 4 P.M. on the 30th, a sudden movement of the ice having opened a narrow channel, we hastened to launch and load the boats; and, pushing them through, succeeded in rounding the Cape. We then ran under sail with a favourable breeze till 11 P.M., when the night being dark we got involved among drift-ice, and not being able to reach the shore dropped anchor off Point Lockyer, and went to sleep in the boats.

We resumed the voyage at 4 A.M. on the 31st, and, getting inside of fields of ice which covered the sea as far as our view extended, we ran along the coast until we came to an island in Basil Hall's

Bay, on which the sea ice rested, barring our further progress on its outside. On the former voyage, this island was thought to be part of the main shore; but on ascending to its summit, which rises about three hundred feet above the sea, we discovered its insular nature, and perceived that the ice within it was not only smoother, but lay less compactly. We therefore took that direction, and found that the inlet runs about five miles behind the island into a narrow valley, bounded by hills between three and four hundred feet high.

In the afternoon we reached Cape Hearne, and ascended its high grounds to look to seaward, from whence we beheld the same impacted floes of ice to which we had of late been accustomed. The cliffs of this cape are composed of a shingly or slaty limestone, and the beach presents much greenish slate-clay, which breaks down like wacké, and becomes brown on the surface, but its relations to the limestone in respect to position could not be made out. The extreme point of the cape is low and sandy; and the country lying immediately to the southward of the limestone ridge, that constitutes the high grounds of the promontory, is flat, grassy, and marshy, forming a fine feeding ground for rein-deer, of which we saw several herds. A considerable stream winds through the plain, and enters the sea about two miles to the southward of the cape. Its mouth, which is barred by a sand-bank, is marked by two cliffs of sand, and it pours out water enough to render the sea clay-coloured for two or three miles, and fresh enough to be drinkable. We encamped, three miles beyond it, on a point formed of slate-clay, of which the beach, after we had passed Cape Hearne, seemed every where to consist. Here we found a decayed sledge, that was put together with copper nails marked with the broad arrow, which must have been extracted from the boats I abandoned on the Coppermine River in 1826.

Since rounding Cape Parry, we had seen very few traces of Eskimos, and had not met a single individual of that nation; but we had now entered a better frequented district, in which traces of the natives abounded. There was a hard frost in the night, with a sharp east-south-east wind blowing from the ice.

The coast being flat, and the water within the ice very shallow, the officers and most of the men walked along the shore, on the morning of the 1st of September, leaving two of the crew in each boat to pole them along. The country is level and swampy, and is crossed by long channels like ditches, on whose banks shale and slate-clay are occasionally exposed. It would seem that on this eastern flank of the limestone formation there is also a shale deposit, but not so extensive a one as that seen on its western side.

In the course of our walk we passed an Eskimo stage, on which, among deer-skins and other effects, we observed the skin of a white bear. We had previously found a skull of this animal on the beach, so that there is no doubt of its frequenting this coast.

After breakfast we made very slow progress, having to cut a way through new ice. It did not exceed an inch in thickness, but, being formed on a foundation of snow, did not crack readily, while, at the same time, it was hard enough to cut the planks of the boats through, rendering them scarcely sea-worthy, though we had strengthened them on the water line with sheets of tin beat out from the pemican cases. In dragging them over the floes they were much shattered. At noon, finding that we could not advance further, in the present condition of the ice, without pulling the boats to pieces and running the risk of losing all our stores and provisions, we encamped about eight miles from Cape Kendall, which bore south-west.

On viewing the sea from the high grounds behind our encampment, and ascertaining that no lanes of open water were visible in any direction, I determined, after consulting with Mr. Rae, to leave the boats at this place, and commence the overland march in the course of two days if no amelioration of the weather or alteration in the state of the sea occurred during the interval. If the weather should improve, it was our intention to remain some days longer, to watch its effects on the ice. The higher grounds at this time were covered with snow, but the lower lands were mostly bare.

The unavoidable conclusion of our sea voyage while still at some distance from the Coppermine River was contemplated by me, and I believe by every individual of the party, with great regret. I had hoped, that by conveying the boats and stores up the Coppermine River beyond the range of the Eskimos we could deposit them in a place of safety to be available for a voyage to Wollaston Land next summer. But abandoned as they must now be on the coast, we could not expect that they would escape the searches of the hunting parties who would follow up our foot-marks, and who were certain to break up the boats to obtain their copper fastenings. The unusual tardiness of the spring, and our unexpected delay on Methy Portage for want of horses, caused our arrival on the arctic coast to be considerably later than I had in secret anticipated, though it differed little from the date I had thought it prudent to mention when asked to fix a probable time. Even a few days, so unimportant in a year's voyage elsewhere, are of vital consequence in a boat navigation to the eastward of Cape Parry, where six weeks of summer is all that can be reckoned upon. Short, however, as the summer proved to be, neither that nor our tardy commencement of the sea voyage would have prevented me from coasting the south shore of Wollaston Land, and examining it carefully, could I have reached it, for the distance to be performed would have been but little increased by doing so. The sole hinderance to my crossing Dolphin and Union Straits was the impracticable condition of the close-packed drift-ice. In wider seas, where fields and large floes exist, these offer a pretty safe retreat for a boat-party in times of pressure, and progress may be made by dragging light boats like ours over them; but the ice that obstructed our way was composed of hummocky pieces, of irregular shape, and consequently ready to revolve if carelessly loaded or trod upon. At certain times of the tide, moreover, they were hustled to and fro with much force.

As only small packs of ice and few in number were seen off the Coppermine by Sir John Franklin in 1820, by myself in 1826, and by Dease and Simpson in 1836 and 1837, being four several summers, the sight of the sea entirely covered so late in August was wholly unexpected, and I attributed so untoward an event to the north-west winds having driven the ice down from the north in the first instance, and to the easterly gales, which afterwards set in, pressing it into that

bight of Coronation Gulf; but Mr. Rae's experience in the summer of 1849 shows that in unfavourable seasons, the boat navigation is closed
for the entire summer, and we learned from a
party of Eskimos whom we met in Back's Inlet, as
I shall have occasion to mention hereafter, that the
pressure of the ice on the coast this summer was
relieved only for a very short time.

The state of the straits produced the melancholy
conviction, that a party, even though provided with
boats, might be detained on Wollaston Land, and
unable to cross to the main; but yet at that time
my apprehensions for the safety of the missing
ships were less excited than they have been since.
For then their absence had not been extended much
beyond the time that their provisions were calculated to last; and, being ignorant of Sir James C.
Ross having been arrested in Barrow's Straits, I
hoped that the accumulation of ice which annoyed
us might be the result of a clearance of the northern channels, and that the two ship expeditions
might have happily met at the very time that we
were no longer able to keep the sea.   It is now
known that the season was equally unfavourable
throughout the arctic seas north of America.

The idea of a cycle of good and bad seasons has
often been mooted by meteorologists, and has frequently recurred to my thoughts when endeavour-

ing to find a reason for the ease with which at some periods of arctic discovery navigators were able to penetrate early in the summer into sounds which subsequent adventurers could not approach, and to connect such facts with the fate of the Discovery ships. But neither the periods assigned, nor the facts adduced to prove them by different writers, have been presented in such a shape as to carry conviction with them, until very recently. Mr. Glaisher, in a paper published in the Philosophical Transactions for 1850, has shown, from eighty years' observations in London and at Greenwich, that groups of warm years alternate with groups of cold ones, in such a way as to render it most probable that the mean annual temperatures rise and fall in a series of eliptical curves, which correspond to periods of about fourteen years; though local or casual disturbing forces cause the means of particular years to rise above the curve or fall below it.

The same laws doubtless operate in North America, producing a similar gradual increase and subsequent decrease of mean heat, in a series of years, though the summits of the curves are not likely to be coincident with, and are very probably opposed to, those of Europe; since the atmospherical currents from the south, which for a period raise the annual temperature of England, must be counterbalanced by currents from the north on other meridians. The annual heat has been diminishing in London ever since 1844, according to Mr. Glaisher's diagram, and will reach its minimum in 1851.

It can be stated only as a conjecture, though by no means an improbable one, that Sir John Franklin entered Lancaster Sound at the close of a group of warm years, when the ice was in the most favourable condition of diminution, and that since then the annual heat has attained its minimum, probably in 1847 or 1848, and may now be increasing again. At all events, it is conceivable that, having pushed on boldly in one of the last of the favourable years of the cycle, the ice, produced in the unfavourable ones which followed, has shut him in, and been found insurmountable; but there remains the hope that if this be the period of rise of the mean heat in that quarter, the zealous and enterprising officers now on his track, will not encounter obstructions equal to those which prevented their skilful and no less enterprising and zealous predecessor in the search, from carrying his ships beyond Cape Leopold.

With respect to the maintenance of a party detained on the islands north of Coronation Gulf, rein-deer and musk-oxen may be procured by skilful hunters; but unless the chase were duly organised, and only the most expert marksmen and

good deer-stalkers suffered to go out, there would
be a danger of the animals migrating from feeding-
grounds on which they were much disturbed.
With nets a large quantity of salmon and other fish
might be captured in Dolphin and Union Straits,
and doubtless also in the various channels sepa-
rating the islands; with percussion guns we had no
difficulty in killing seals, and we might, had we
chosen, have slain hundreds, though, as they dive at
the flash, the chance of shooting them with a ship's
musket having an ordinary lock, would be greatly
diminished. Swans, snow geese, brent geese, eiders,
king ducks, cacawees, and several other waterfowl,
breed in immense numbers on the islands; and the
old ones when moulting, and the young before they
are fledged, fall an easy prey to a swift runner, and
still more surely to a party hemming them in and
cutting off their retreat.

To people acquainted with the Eskimo methods of
building ice and snow houses, shelter may be raised
on the bleakest coast, except in the autumn months;
but, unless blubber were used as fuel, there would
be a difficulty in maintaining fire for cooking by
any one who has not the genius for turning every
thing to account which Mr. Rae evinced, when he
boldly adventured on wintering on a coast bearing
the ominous appellation of Repulse Bay, with no
other fuel than the *Andromeda tetragona*, — an inte-

resting and beautiful herb in the eye of a botanist, but giving no promise to an ordinary observer that it could supply warmth to a large party during a long arctic winter. To apply it, or any of the other polar plants, to such a purpose, a large quantity must be stored up near the winter station before the snow falls.

I have thought it right to throw these few observations together in this place, that a reader unacquainted with the natural resources of the country may judge of the probability of the whole or part of the crews of the Erebus and Terror maintaining themselves there, supposing the ships to have been wrecked. Of course, as long as the vessels remained, they would afford shelter and fuel; but the other contingencies would come into consideration if parties went off in various directions in quest of food. One great purpose of the expedition which I conducted along the coast was to afford relief to such detached parties, or to the entire crews, had they directed their way to the continent, and our researches proved at least that none of the party, having gained that coast, were dragging out a miserable existence among the Eskimos, without the means of repairing to the fur posts. In the following summer of 1849 Mr. Rae ascertained that the Eskimo inhabitants of Wollaston Land had seen neither the ships nor white men. The knowledge of these facts had an influence with the Admiralty in concentrating the future search in the vicinity of Melville Island; Captain Collinson and Commander Pullen being directed to approach its coasts from the westward, while Captain Austen, and the squadron of hardy navigators in his wake, were to trace the Discovery ships from the eastward. A more ample and noble effort to rescue a lost party was never made by any nation, and it has been humanely seconded from the United States of America. May God bless their endeavours!

CHAP. X.

PREPARING FOR THE MARCH. — SLEEP IN BACK'S INLET. —
ESKIMO VILLAGE. — ESKIMOS FERRY THE PARTY ACROSS RAE
RIVER. — BASALTIC CLIFFS. — CROSS RICHARDSON'S RIVER. —
MARCH ALONG THE BANKS OF THE COPPERMINE. — GEESE. —
FIRST CLUMP OF TREES. — MUSK-OXEN. — COPPER ORES AND
NATIVE COPPER. — KENDALL RIVER. — MAKE A RAFT. — FOG.
— PASS A NIGHT ON A NAKED ROCK WITHOUT FUEL. —
FINE CLUMP OF SPRUCE FIRS. — DISMAL LAKES. — INDIANS. —
DEASE RIVER. — FORT CONFIDENCE. — SEND OFF DESPATCHES
AND LETTERS.

ON the 1st and 2nd of September we had northerly
and north-east winds, with a low temperature, sleet,
snow, and occasional fogs.   We were all employed
in preparing the packages for the march, consisting
of thirteen days' provision of pemican, cooking
utensils, bedding, snow shoes, astronomical instruments, books, ammunition, fowling-pieces, portable
boat, nets, lines, and a parcel of dried plants.
These were distributed by lot, each load being
calculated to weigh about sixty or seventy pounds.
Mr. Rae voluntarily resolved to transport a package
nearly equal to the men's in weight; but, distrust-
ing my own powers of march, I made no attempt
at carrying such a load as I had done on a former
voyage, and restricted myself to a fowling-piece,

ammunition, a few books, and other things which I thrust into my pockets. Six pieces of pemican were buried under a limestone cliff, together with a boat's magazine full of powder. The tents were left standing near the boats, and a few cooking utensils and hatchets deposited in them for the use of the Eskimos.

After an early breakfast on the morning of Sunday the 3rd of September we read prayers, and then set out at six o'clock. At first we pursued a straight course for the bottom of Back's Inlet, distant about twelve geographical miles; but finding that we were led over the shoulder of a range of hills on which the snow was deep, we held more to the eastward, through an uneven swampy country, where we saw many deer feeding; but made no attempt to pursue them.

The men, with a few exceptions, walked badly, particularly the two senior seamen, and after we had gone a few miles, were glad to lighten their loads by leaving their carbines behind. At half-past three we reached the inlet, about seven miles from the pitch of Cape Kendall, and halted for the night under a cliff of basalt two hundred feet high. The inlet and the sea in the offing were full of ice, and the weather continued cold; but some scraps of drift-wood, chiefly willows, being found on the beach, we managed to cook supper; and, selecting the best sleeping places we could find among the blocks of basalt, passed a pretty comfortable night.

We started a quarter before six on the morning of the 4th, to walk round the inlet; and Frazer having sprained his knee on the preceding day, we were constrained to lighten his load by leaving a large hatchet, and distributing a portion of his pemican among the others. Our course along the inlet was south 74° west for four miles and a half, when we perceived ten Eskimo tents on the opposite shore. Mr. Rae and Albert went ahead of the men, who were straggling very slowly along; and on coming opposite to the tents, and shouting, three Eskimos crossed the inlet in their kaiyaks, and cordially consented to ferry the whole party over. This small tribe have no " umiaks;" and, as the kaiyaks carry only one person, some contrivance was requisite to render them available as ferry-boats. Our friends had already learned how to effect this from their intercourse with Mr. Simpson and his party in 1838, viz., by placing two poles across a pair of kaiyaks, and lashing them firmly together. In this way a single paddler could take over a sitter and his bundle. Four kaiyaks, being all they could muster, were brought into requisition, by which, with the addition of Lieutenant Halkett's portable boat, three men with their loads could be ferried over at each trip. At the place where we crossed, the inlet had contracted to the breadth of four hundred yards ; and is there, in fact, a river, since its water is fresh.   The whole party was landed on the southern shore by eleven o'clock. On the river I bestowed the name of my active, zealous, and intelligent companion Mr. Rae, as a testimony of my high sense of his merits and exertions, which had been called forth to the uttermost in our late endeavours to push on through the ice. It was mainly through his skill and perseverance that we had been enabled to travel as far as we did by sea, and thus shorten the land journey ; which, with an increased distance, and, consequently, proportionably augmented loads, would have been a very arduous undertaking indeed to some of our party. We considered ourselves as very fortunate in obtaining the assistance of a friendly party of Eskimos at this place, on learning from them that the river kept its width, and was not fordable for a long way up the country.   Mr. Rae, in the succeeding spring, ascended it for twenty miles, and ascertained that it flowed directly from the west, and was about the size of the Dease, or about one hundred and twenty yards wide.   Its bed is limestone; and a range of basaltic cliffs, varying from fifty to two hundred feet in height, skirts its northern bank.   These cliffs are a continuation of the magnificent precipices, which, commencing at Cape Kendall, rise at intervals

x 4

of three or four miles on the north shore of Back's
Inlet, their faces being to the southward, and their
line of direction or strike nearly due east and west.
At Cape Kendall the basalt is obscurely columnar,
and rests on a bed of compact felspar, containing
minute grains of a green mineral.　At a cascade
in Rae River, ten miles above its mouth, walls
from eight to twenty feet high, of bluish-grey quartz
rock in thin layers, hem in the stream.　Salmon and
other fish ascend a shelving shoot of the cascade.　At
this place Mr. Rae discovered, among the limestone
and quartz rock, layers of asparagus-stone, or
apatite (phosphate of lime), thin beds of soap-stone,
and some nephrite, or jade,—a group of minerals
which belongs to primitive formations; and from
the similarity of the various rocks associated in this
quarter to those occurring at Pigeon River, and
other parts on the north shore of Lake Superior, I
am inclined to consider that the two deposits
belong to the same geological era, both being more
ancient than the silurian series.　Neither Mr. Rae
nor I discovered any organic remains in the limestone.

　Among the Eskimos here encamped we recog-
nised two mentioned by Mr. Simpson, one having a
wen on his forehead, and the other being a very old
man who walked on crutches.　The kind treatment
and presents they received from Messrs. Dease and

Simpson had impressed them with a favourable opinion of the dispositions of white men, and doubtless was the cause of their readiness to come to our assistance, and to put themselves and their families so completely in our power. Our men bought sealskin boots from them, which proved very useful; and we paid the man with the wen, who was the leader, for his services in ferrying us over, with two hatchets, which were of great value to him. I had cautioned every one against offering this harmless, good-natured people any offence; and I must give our men the credit of having strictly adhered to the orders they received. I believe I was the only one who entered any of their huts; and I did so for the purpose of presenting some needles and other articles to the women, and obtaining a glimpse of their ménage. In one tent six or seven women were seated in a circle sewing. They were nearly naked, very dirty, hung their heads down, and seemed to be much afraid. As the females we met on the coast, who showed neither fear nor shamefacedness, were generally clean, I believe that the apprehensions of these poor women had caused them to rub ashes or mud on their faces and persons. They received my presents, but seemed to be relieved when I took my leave. Before we quitted the encampment, several younger men joined from the northern shore of the inlet; and we

learnt that we had interrupted their day's occu-
pation in killing rein-deer.   The more active among
them go at this season to the meadows which we
had crossed on the previous day, and gradually drive
the animals to the inlet, hemming them in, and
compelling them, with the aid of their dogs, to take
the water.   As soon as this takes place, the rest of
the party, who are lying in wait in their kaiyaks,
paddle towards the herd, and spear as many of
them as they can.   A considerable quantity of deer's
meat was hanging to dry on stages; and we purchased a little of it for our evening meal.

These people told us, as I have mentioned already, that the ice had parted from the shore only
a very short time this season, which, they added, was
almost unprecedented within their recollection.
Their migrations extend only to the lower part of
the Coppermine River on one side, and a short way
along the coast on the other.   They communicate
occasionally with the Eskimos of Wollaston Land,
but none of them had been so far to the westward
as the sources of Rae's River.   The want of umiaks
was a sufficient indication of the shortness of their
migrations seaward.

Our friend with the wen accompanied us three
or four miles on our journey, to show us a ford
across Richardson's River; but the number of questions he put to Albert respecting the boats, showed

that his thoughts were directed to the treasures he expected to find in them; and at length he turned back, after pointing out the direction in which we ought to go. Albert had been told not to mention the place where the boats were left; but the Eskimos could without the slightest difficulty trace up the foot-marks of so large a party as ours; and I believe that by the evening, or early next day, most of the party were assembled in our deserted tents.

We arrived on the banks of Richardson's River about three o'clock, but failed in finding a ford; and, the walking being bad, some of the men lagged far behind, which induced us to encamp early. Richardson's River, as well as Rae's, is flanked by lofty precipices of basalt, which, coming successively into view, produce striking vistas in a bleak and otherwise uninteresting country. From the summit of one of these eminences near our encampment I obtained a wide view of the land, and saw a line of cliffs running along the Rae from Cape Kendall; another rank marks out the course of the Richardson, from Point Mackenzie* up to the junction of its two branches, where the cliffs also fork off at an acute angle, a series of them skirting the valley of each branch. A range of cliffs, but

---

* At this point the basalt is superimposed on a dark bluish-grey crystalline limestone.

of a less imposing character, forms the western
boundary of the valley of the Coppermine, separat-
ing it from that of the Richardson. All these
rows of precipices face towards the south, south-
east, or east-south-east, and radiate between west
and south-south-west from a point in Coronation
Gulf, at which they would meet if prolonged. The
western boundary of the granite formation appears
in the islands of that gulf, associated with many
trap rocks; in the form of lofty hills at Cape Barrow; again at the bend of the Coppermine, on the
south side of Kendall's River; on the north-east and
eastern arms of Great Bear Lake; on Point Lake;
in country round Fort Enterprise; and from thence
to Fort Providence and across Great Slave Lake to
the mouth of Slave River, and so onward to Athabasca.

Richardson River was discovered in 1822 by some
hunters of Sir John Franklin's party, and, on their
report, it received its present name from that
officer; but its outlet was erroneously supposed to
be only four or five miles to the west of the Coppermine. In 1826, I ascertained that its supposed
mouth was only a shallow bay; and, in 1838–9,
Mr. Simpson examined the river, and proved that
it falls into Back's Inlet; on which occasion he
confirmed the appellation which Sir John Franklin had given it. Its junction with the inlet was

ascertained by Mr. Simpson to be in lat. 67° 53′ 57″ N., long. 115° 56′ W.

Commencing the day's journey at six in the morning of the 5th, we crossed, about an hour afterwards, a small tributary of the Richardson; and at nine, having then walked about four miles and a half from our sleeping place without discovering a ford, we determined on crossing in Lieutenant Halkett's boat, though, as it could carry no more than two men at a time, the operation was likely to be tedious. Some tall willows (seven or eight feet high), growing on the muddy banks of the river, afforded us the means of making a fire and preparing breakfast. In the mean time, all the net lines, spare lines, and carrying slings were united to form a hawser, wherewith we might draw the boat backwards and forwards. Mr. Rae and Albert crossed first; and, owing to the man to whom the paddles had been assigned as part of his load having left them behind, they had no other instruments for propelling the boat than two tin dinner-plates. They succeeded, however, in crossing, though their hands were much chilled by the ice-cold water; and subsequently the whole party were drawn across. The width of the stream, by measuring half the line, was ascertained to be one hundred and forty yards.

At one o'clock all had crossed; and, the bundles

being again duly arranged, we resumed the march, and in a short time gained the summit of the ridge dividing the valley of the Richardson from that of the Coppermine. The latter was clothed with snow, the climate being seemingly more severe, though the distance between the streams is so small. The plain which lay at our feet, between the ridge and the Coppermine, is so much inter-sected by small lakes, that we chose the driest line of march, rather than the most direct, to avoid the necessity of fording the lakes, or losing ground by rounding them. At three o'clock we reached the banks of the river, three or four miles above Bloody Fall; and, having found a sufficiency of wood, made a good fire, which of late had been a very rare luxury. Many deer, Hutchin's and snow geese were seen; and, Mr. Rae having killed nine or ten of the latter, we enjoyed an excellent supper.

The country within the influence of the sea-breezes which come from the icy surface of Coro-nation Gulf, has the barren aspect and poor climate of the *tundras* of the Siberian arctic region. The moister tracts, where the soil is clayey, retain so much ice-cold water in the short summer, that a sparing vegetation exists only in the hassocks, which bear, among the Chepewyan tribes, the name of " women's heads," and render the footing of pedestrians insecure and dangerous. " You may kick them," say the ungallant Indians, " but they cause you to stumble and never go out of the way." In the drier, sandy, and gravelly spots, which are more common among the primitive rocks, the ground is covered with the lichens on which the musk oxen and rein-deer feed. Of these the *corniculariæ* and *cetrariæ* are the most important ; and they are most prized by the animals when the melting snow in spring renders them soft and tender. As the season advances, the grasses and bents which flourish in sheltered valleys furnish the chief food of the herbivorous animals; and, when the snows fall, the rein-deer retreat southward to the woody districts, into which they penetrate deeper in severe weather, and in the milder intervals return to the barren grounds to scrape the hay from beneath the snow. The suddenness of the winter in these high latitudes serves the important purpose of arresting the juices of the grasses and freezing them, so that until late in spring they retain their seeds and nutritive qualities without withering. It has the same effect on the berry-bearing plants. The crow-berry (*Empetrum nigrum*), bleaberry (*Vaccinium uliginosum*), and cranberry (*Vaccinium vitis idea*), which grow in profusion among the lichens of the arctic wastes, not only furnish fruits for the bears and geese in autumn, but retain them in perfection until the ground begins to dry up under the influence of the hot summer suns, and the new flowers are expanding. In the month of September the snow-geese (*Anser hyperboreus*), and Hutchin's geese (*Anser Hutchinsii*), feed much on the crow-berries, which render them fat and well-flavoured. The first-named geese breed in Wollaston Land, to which they cross in the beginning of June. We had noticed, while on the coast of Dolphin and Union Straits, the earliest bands travelling southwards again in the middle of August, so that their stay in their native place falls short of three months. The Hutchin's geese and brent geese breed on the coasts of the Arctic Sea, and the laughing geese (*Anas albifrons*) resort to the country north of the Yukon, beyond the arctic circle. The Canada geese, or "bustards" of the Canadians (*les outardes*), breed throughout the woody districts, but do not reach the vicinity of the Arctic Sea, except on the banks of some of the large rivers. The most northern localities in which we observed them were the channels between the alluvial islands which form the delta of the Mackenzie.

On the 6th we had clear weather with a hard frost, and gladly welcomed the face of the sun, which had been a stranger to us for more than a fortnight. The swamps being frozen over so as to support a man's weight, the party generally walked more briskly than usual; but three of the seamen and two of the sappers and miners were so lame, that we were obliged to make long and frequent halts to allow them to close in, and were unable to accomplish two geographical miles in the hour. To spare their strength, we encamped at the early hour of 2 P.M., having marched about ten miles and a half. Deer, geese, and ptarmigan, were seen in abundance during the day. In the evening the weather became cold, with rain, snow, and hail.

On the 7th our morning's march was performed in a snow-storm, with a chilly northerly wind. About four miles from last night's sleeping-place, we came to a chain of narrow lakes, lying parallel to the river, and emptying themselves into it by a small stream which issues from their northern extremity. They are three miles in length, and lie about a mile from the river. We afterwards forded two rapid torrents full of large greenstone boulders. One of them flows through a narrow chasm in friable dark-red sandstone, and the other is bounded by cliffs of red quartz rock, or perhaps of trap, but I could not approach them near enough for examination. The discomfort of the march was greatly augmented by the freezing of our clothes, wet in crossing the streams, and we

gladly encamped, at two o'clock, on coming to a
clump of stunted white spruce trees, where we
arranged a comfortable bivouack, by placing small
branches between the frozen ground and our
blankets.   In the existence of many scattered
stumps of decayed spruce fir trees, and the total
absence of young plants, one might be led to infer,
that of late years the climate had deteriorated, and
that the country was no longer capable of support-
ing trees so near the sea-coast as it had formerly
done.   Many plants of different species of *Pyrola*
grow on the sea-shore; and as these are most
abundant in forest lands, it is possible that they
may be the memorials of ancient woods.   The
largest tree in the clump in which we bivouacked
had a circumference of thirty-seven inches at the
height of four feet from the ground.   Its annual
layers were very numerous and fine, and indicated
centuries of growth, but I was unable to reckon
them.   This place lies in lat. 67° 22′ N.

The evening proving fine, Mr. Rae and Albert
went out to hunt, and both had the pleasure of
seeing the musk-ox, for the first time in their lives.
The *uming-mak* is known by name and reputation
to all the Eskimo tribes; but as it does not exist
in Greenland, or Labrador, nor in the chain of
islands extending north from that peninsula along
the west side of Davis Straits, Albert, who was a

native of East Main, now for the first time approached its haunts. Mr. Rae, with the feelings of an ardent sportsman, had longed to encounter so redoubtable an animal; and the following is an account of the meeting :—

On perceiving a herd of cows, under the presidency of an old bull, grazing quietly at the distance of a few miles from our bivouack, he and Albert crept towards them from to leeward; but the plain containing neither rock nor tree behind which they could shelter themselves, they were perceived by the bull before they could get within gun-shot. The shaggy patriarch advanced before the cows, which threw themselves into a circular group, and, lowering his shot-proof forehead so as to cover his body, came slowly forwards, stamping and pawing the ground with his fore-feet, bellowing, and showing an evident disposition for fight, while he tainted the atmosphere with the strong musky odour of his body. Neither of the sportsmen were inclined to irritate their bold and formidable opponent by firing, as long as he offered no vital part to their aim; but, having screwed the bayonets to their fowling-pieces, they advanced warily, relying on each other for support. The cows, in the meantime, beat a retreat, and the bull soon afterwards turned; on which Mr. Rae fired, and hit him in the hind quarters. He instantly faced about, roared, struck the ground forcibly with his fore feet, and seemed to be hesitating whether to charge or not.   Our sportsmen drew themselves up for the expected shock, and were by no means sorry when he again wheeled round, and was, in a few seconds, seen climbing a steep and snow-clad mountain side, in the rear of his musky kine.

These animals inhabit the hilly, barren grounds, between the Welcome and the Copper Mountains, from the sixty-third or sixty-fourth parallels to the Arctic Sea, and northwards to Parry's Islands, or as far as European research has yet extended. They travel from place to place in search of pasture, but do not penetrate deep into the wooded districts, and are able to procure food in winter on the steep sides of hills which are laid bare by the winds, and up which they climb with an agility which their massive aspect would lead one ignorant of their habits to suppose them to be totally incapable of.   In size they are nearly equal to the smallest Highland or Orkney *kyloes;* but they are more compactly made, and the shaggy hair of their flanks almost touches the ground. In structure they differ from the domestic ox, in the shortness and strength of the bones of the neck, and length of the dorsal processes which support the ponderous head.   The swelling bases of the horns spread over the foreheads of both sexes, but are most largely developed in the old males. The musk-ox has also the peculiarity in the bovine tribe of the want of a tail; the caudal vertebræ, only six in number, being very flat, and nearly as short, in reference to the pelvis, as in the human species; the extreme one ending evenly with the tuberosities of the ischium. A tail is not needed by this animal, as in its elevated summer haunts moschetoes and other winged pests are comparatively few, while its close, woolly, and shaggy hair furnishes its body with sufficient protection from their assaults. The fore-pasterns are provided on their outsides with a slender accessory bone, of about half their length. The fossil Irish elk and musk-deer have also rudimentary toes, but of a different form. Though I have not been able to ascertain that the range of the species was ever greater than it is known to be at present, I have read somewhere of a skull having been found in Greenland. One in tolerable preservation, but defective in the nose, was procured by Captain Beachey, from that very curious deposit of bones in the frozen cliffs of Eschscholtz Bay of Beering's Straits. That skull is now preserved in the British Museum, and a perfect skeleton of the recent animal exists in the museum at Haslar Hospital.

Sept. 8th. — A meridional observation was ob-

tained to-day in lat. 67° 17′ N.   We crossed a projection of the Copper Mountains, to cut off a considerable bend of the river; and, at four in the
evening, reached its banks again, and encamped.
While among the hills we had to walk in snow
shoes, with much fatigue; but in the afternoon a
thaw took place in the low grounds, under the
influence of a warm sun; and we were annoyed
by sand-flies in the evening.   I noticed that the
upper branches of the scrubby spruce firs, among
which we encamped, were confined to their south-
east and southern aspects.   The lower branches,
as usual in such exposed situations, lay close to the
ground, and spread widely, considering the small
height of the tree.

The effect of the last two or three days' march
proved to me that I had over-calculated my strength,
in loading and clothing myself too heavily.   I
therefore transferred my gun and part of my
clothing to Dore, an active young seaman, who
was always at the head of the line, and whose load,
as well as that of the others, had been reduced by
the consumption of pemican.   Some of the worst
walkers had already been eased of everything but
their blankets, spare clothing, and a few pounds of
pemican, but they still lagged in the rear.

In this neighbourhood, in 1826, we found a vein
containing malachite and other ores of copper, with

some of the native metal scattered in detached pieces. The Indians procure the metal on both sides of the Coppermine, in a district which requires several days to traverse. A rolled piece of chromate of iron was picked up on the banks of the river by Mr. Rae. This mineral, so valuable on account of the beautiful pigments which are manufactured from it, is found, according to Jameson, in primitive porphyry, and in beds between clay-porphyry and wacké, and more abundantly in America than on the Old Continent.

The 9th proved to be another fine day. Commencing our march a little before six, we halted at noon for an hour and a half, and encamped at five. A meridional observation gave the lat. 67° 14′ 32″ N. In the afternoon we passed the boat left by Dease and Simpson in 1839, which required too much repair to render it water-tight, or we should have availed ourselves of it for the remainder of the river course we had to follow.

Starting at the usual hour on the 10th, we struck the Kendall about a mile and a half from its junction with the Coppermine, after a march of five hours and a half. Mr. Rae went down to its mouth to look for a note which we expected to find, as I had directed James Hope, with two or three Indians, to meet me there; or, if he arrived earlier than us, to leave a memorandum and descend the river as far as Bloody Fall. This arrangement, which was made in anticipation of our bringing the boats up the river, was my chief reason for making the circuit of the Coppermine; for our most direct course, after leaving Back's Inlet, would have been by tracing up Richardson River, and crossing the mountains more nearly in the parallel of Fort Confidence. As we had discovered no foot-marks of the party on our march, we concluded that they had not arrived; and Mr. Rae confirmed this opinion by his report of the absence of any signal mark at the mouth of the Kendall. From specimens of the rocks obtained by this gentleman, I ascertained that the walls of the gorge by which the stream enters the Coppermine are composed of red quartz rock disposed in thin layers. The mouth of the Kendall is laid down by Mr. Simpson in lat. 67° 7′ N., long. 116° 21′ W.; and a meridional observation gave 67° 06′ 43″ N., as the latitude of the place where we fell upon the stream.

We walked for nearly three miles along its banks to look for a crossing-place; but, finding that it was no where fordable, we resolved to construct a raft, as there was a sufficiency of dry timber for the purpose. We therefore encamped, and Mr. Rae superintended the operation of raft-making. The weather being mild we were again troubled with sand-flies.

Sept. 11*th.* — During a fine night we enjoyed the light of a full moon; but towards the morning the wind veered to the north-west, and a moist, chilling fog enveloped us. Our raft could support three at a time, and enabled us all to cross by seven o'clock. A fresh arrangement of the loads was made here; and, to lighten them as much as possible, I deposited my packet of dried plants and some books in a tree, intending to send for them in the winter. After breaking up the raft to recover the lines by which we had fastened it, we piled the logs up on the bank to attract the attention of Hope's party, should we happen to miss them.

Our course was shaped directly across the country for Dease's River; and as we ascended the high grounds the fog became more dense, so that by noon we could not see beyond two or three yards. We steered by the compass, Mr. Rae leading, and the rest following in Indian file. I kept rather in the rear to pick up stragglers; but, though we walked at a much brisker pace than usual, there was little loitering. The danger of losing the party made the worst walkers press forward. On the hills the snow covered the ground thickly; and it is impossible to imagine any thing having a more dreary aspect than the lakes which frequently barred our way. We did not see them until we came suddenly to the brink of the rocks which

bounded them, and the contrast of the dark surface
of their waters with the unbroken snow of their
borders, combined with the loss of all definite outline in the fog, caused them to resemble hideous
pits sinking to an unknown depth.   The country
over which we travelled is composed chiefly of
granite ; and after walking till half-past five without perceiving a single tree, or the slightest shelter,
we came to a convex rock, from which the snow
had been swept by the wind.   On this we resolved
to spread our blankets, as it was just big enough
to accommodate the party.   There being no fuel of
any kind on the spot, we went supperless to bed.
Some of the party had no rest, and we heard them
groaning bitterly ; but others, among whom were
Mr. Rae and I, slept well.   We learned afterwards
that a clump of wood grew within a mile and
a half of our bivouack ; but even had we been
apprised of its existence, we could scarcely have
found it in the fog.   Several showers of snow
occurred in the day, and some fell in the night.

Had it not been for the fog, we should have met
James Hope and two Indians this day, for they
were not many miles distant in the morning ;
but, notwithstanding their acquaintance with the
country, they went astray in the thick weather,
and did not reach the place where we crossed
the Kendall till the second day afterwards.

Perceiving then by the remains of the raft that we had crossed, they traced our foot-marks, and, following with their utmost speed, reached our bivouack on the rock two days after we left it.

Commencing the day's march at half-past four in the morning of the 12th, we came to a tributary of the Kendall at eight. In fording this, the water came up to our waists, and we were all more or less benumbed; but a few trees on the bank furnished us with the means of making good fires; and by the time that we had finished breakfast we were comfortably dry. A meridional observation gave us lat. 67° 09′ N.

At two we came to another branch of the Kendall, which runs through a ravine of red and spotted sandstone, under whose shelter there grew a remarkably fine grove of white spruces. The best-grown tree measured sixty-three inches in circumference, and did not taper perceptibly for twenty feet from its root. Its total height was from forty to fifty feet. Other trees of equal girth tapered more, and one decayed trunk, which lay on the ground, looked to be considerably thicker. We encamped in this snug place, and Mr. Rae and Albert, employing the evening in the chase, killed a rein-deer and some snow geese.

Mr. Rae endeavoured in the winter to measure the height of the creek on which we encamped this night, and of other remarkable places on the route between Great Bear Lake and the Coppermine River, by the aneroid barometer; but that instrument during the journeys underwent such a change, that no reliance could be placed on its indications, when they were compared with those of the barometer at Fort Confidence. The same inconvenience, however, did not materially affect observations made on it at short intervals of time; and in this way the brow of the hill to the south of the creek was ascertained to be six hundred and seventy feet above the stream.

Onwards from the level of this brow the country is a gently undulated plain, which is bounded on the south at the distance of a few miles by an even range of hills two or three hundred feet high, and far to the north by the Coppermine Hills, which Lieutenant Kendall and I crossed in 1826, as mentioned in the narrative of Sir John Franklin's second overland journey. A range of lakes, named by Mr. Simpson the Dismal Lakes, lies between these hills and our line of route. They are skirted by broken belts of wood, but the rest of the country is quite naked, the few dwarf trees that exist on the plain being concealed in the depressions of the water-courses of the small rivulets.

The comfortable supper of venison, a sound night's rest in an encampment where nothing was wanting, and the lighter loads, had such an effect on the spirits of the party, that we mounted the hill above the ravine on the morning of the 13th with unusual alacrity, and kept together in close single file. Travelling in this way, our line, as it undulated over the gentle swellings of the plain, was seen from afar, and we were discovered very soon after emerging from the ravine by a party of Indians, encamped on the side of a hill about six miles distant. Happily for these people they knew we were now on the march, and expected to see us at this time; for had it not been so they would have fled instantly with their wonted timidity, and most probably have left every thing they possessed behind them. As it was, we were not many minutes in sight before they signalled their position by raising a column of smoke. This was replied to by us as soon as we could strike a light and gather a few handfuls of moss; and our answer was immediately acknowledged by them with a fresh column. They were encamped nearly at right angles to our line of route; but I thought it better to join them for the purpose of obtaining intelligence, and we accordingly struck off in that direction.

We reached their tents a little before noon, time enough for us to make a meridional observation, by which we ascertained that the latitude was

67° 11′ 30″ N., and the sun's bearing at noon
S. 50° E.   These Indians informed us that James
Hope and his companions had been with them five
days previously, and that he had then been two
days absent from the fort.

The site of their encampment was selected for
the commanding view it possessed of the neighbouring country, so that they could mark the
movements of the herds of rein-deer and musk-
oxen that at this season were numerous.   Their
chase was successful, and their condition and that
of their dogs showed that they were revelling in
abundance.   No doubt this party might now have
laid up a sufficiency of venison to feed them,
with due economy, all the winter; but such is not
the habit of the nation.   When the pressure of
want ceases to be felt their exertions flag, and
they consider it useless to store up provision which,
according to their custom, is at the mercy of every
idle and hungry person of the tribe.

They gladly sold us some meat for ammunition,
and would readily have parted with their whole
stock on hand, but I had no desire to load my
party again.   We agreed, however, with one of the
young men to accompany us to the fort, that he
might lead us by the best paths, and waited for an
hour until he had prepared a heavy load of half-
dried meat, to carry with him as an article of trade.

In the afternoon our way lay over hills of spotted grey sandstone, sandy shale beds, and towards the evening over knolls of gravel. The day's journey was seventeen geographical miles.

Our march on the 14th was made in a south-west by west direction, and was short, for our guide complained of being fatigued by his load. We relieved him of a part, by distributing about forty pounds of it among the men for their supper.

The country we crossed in the course of the day is composed of sandstone, with gravel banks, and undulates, but is not mountainous. Thin groves of trees occur here and there, especially on the borders of rivulets, and many dwarfish and ancient dead stumps remain on the sides of the eminences. The soil is cracked, hummocky, and swampy, and affords uneasy footing to pedestrians. I found much comfort by walking immediately behind the Indian, that I might avail myself of his quick eye, and tread exactly in his footsteps.

We set out early on the 15th, that we might reach the fort betimes. We lost, however, a considerable time, while the guide went in pursuit of several bands of deer that crossed the path. His skill in hunting was indifferent, and he had no success. The morning was snowy. Before noon we forded a branch of the Dease, and at two o'clock came to the banks of that river at the first

rapid.   Here we found a barge moored for our use,
and, embarking the whole party in it, reached the
house at 4 P. M. We were happy to find Mr.
Bell and his people well, and the buildings much
further advanced than we had expected.   All the
houses erected by Dease and Simpson had been
burnt down, except part of the men's dwelling.
Mr. Bell reached the site on the 17th of August, and
immediately set to work.   Since that time he had
built an ample storehouse, two houses for the men,
and a dwelling house for the officers, consisting of
a hall, three sleeping apartments, and store-closet.
This building was roofed in when we arrived, but
the flooring and ceiling of the rooms were not yet
laid, though planks had been sawn for that purpose;
the kitchen was still to be built, and tables, chairs,
and other articles of furniture, to be made.   In the
log houses, which are commonly erected in this
country, the chimneys are massive affairs of tempered clay and boulder stones, and require to be
leisurely constructed.   The Canadians, who are all
practised in the use of the axe, soon set up the wood-
work; and Bruce, the guide, who superintended the
operations, and indeed did two men's work himself,
advanced them rapidly.

Mr. Bell and Mr. Rae quartered themselves with
Bruce in the store-room, and I took possession of
my sleeping room, which was put temporarily in

order. I could there enjoy the luxury of a fire while I was preparing despatches for the Admiralty and writing my domestic letters, though the walls not being as yet clayed, the snow drifted in between every log. The 16th of September was employed in writing, and on the 17th, being Sunday, we assembled in the hall, where I read divine service and returned thanks to the Almighty for our safety. The fishermen who were stationed about five miles from the house came in on this day, so that the whole party were met together. The Canadians, though Roman Catholics, were present on the occasion ; and most of them regularly attended our Sunday services in the winter. In addition to the party from the coast, Mr. Bell had with him here fourteen men, with three women and four children ; so that we had in all forty-two souls to provide for, exclusive of Indians coming casually on our store.

On Monday, the 18th of September, the packet of letters was placed in charge of François Chartier and Louis La Ronde, who were directed to carry it on without delay to Isle à la Crosse, where the wife of the latter resided. Henry Smith, Josephe Plante, and Henry Wilson, Canadians, accompanied them for the purpose of wintering at the fishery on Big Island, Great Slave Lake ; and with them I sent the following men of the English party, whose services could be well dispensed with at our winter quarters :

Stairs, Sully, and Clarke, *seamen ;* Frazer, Dall, Dodd, Sulter, Hobbs, Ralph, Geddes, Webb, Weddell, and Bugbee, *sappers and miners.* Being thus relieved from the maintenance of eighteen people, the resources of the post were considered equal to feeding the remainder, and I looked forward to the winter without anxiety.

Mr. Bell had placed two fishermen, by my desire, at the west end of Great Bear Lake, near its outlet, to be ready to feed my party, had I found it necessary to return up the Mackenzie. I judged it prudent to continue these men there, not only as their fishing hut would be a convenient station for parties travelling to and fro, between Fort Confidence and the posts on the Mackenzie, but also that they might give aid, should our fisheries near the fort fail.

CHAP. XI.

ON THE ESKIMOS OR INUIT.

THE FOUR ABORIGINAL NATIONS SEEN BY THE EXPEDITION. — ESKIMOS. — ORIGIN OF THE NAME. — NATIONAL NAME *INU-IT*. — GREAT EXTENT OF THEIR COUNTRY. — PERSONAL APPEARANCE. — OCCUPATIONS. — PROVIDENT OF THE FUTURE. — VILLAGES. — SEAL HUNT. — SNOW HOUSES. — WANDERINGS NOT EXTENSIVE. — RESPECT FOR TERRITORIAL RIGHTS. — DEXTEROUS THIEVES. — COURAGE. — TRAFFIC. — COMPARED TO THE PHŒNICIANS. — SKRELLINGS. — WESTERN TRIBES PIERCE THE LIP AND NOSE. — FEMALE TOILET. — MIMICS. — MODE OF DEFYING THEIR ENEMIES. — DRESS. — BOATS. — KAIYAKS. — UMIAKS. — DOGS. — RELIGION. — SHAMANISM. — SUSCEPTIBILITY OF CULTIVATION. — ORIGIN. — LANGUAGE. — WESTERN TRIBES OF THE ESKIMO STOCK. — TCHUGATCHIH. — KUSKUTCHEWAK. — A KASHIM OR COUNCIL HOUSE. — FEASTS. — QUARRELS. — WARS. — CUSTOMS. — MAMMOTH'S TUSKS. — NATIONAL NAMES. — NAMOLLOS OR SEDENTARY TCHUTCHKI. — REIN-DEER TCHUKTCHE. — THEIR HERDS. — COMMERCE. — SHAMANISM. — OF THE MONGOLIAN STOCK.

To keep the interruptions of the narrative within reasonable limits, I have hitherto avoided saying much of the native tribes that occupy the countries through which the Expedition travelled, and shall here supply that deficiency by giving some details of the manners and customs of the four nations whose boundaries we crossed in succession.

Reversing the order of our journey, the first of

the native nations that presents itself in descending from the north, is that of the *Eskimos*, as Europeans term them. This appellation is probably of Canadian origin, and the word, which in French orthography is written *Esquimaux*, was probably originally *Ceux qui miaux* (*miaulent*), and was expressive of the shouts of *Tey-mō*, proceeding from the fleets of kaiyaks, that surround a trading vessel in the Straits of Hudson, or coasts of Labrador. The sailors of the Hudson's Bay Company's ships, and the Orkney men in the employment of the Company, still call them *Sŭckĕmŏs* or *Seymŏs*. Some writers, however, have thought the word to be a corruption of the Abenaki term *Eskimantik*, signifying " eaters of raw flesh," which is certainly a habit peculiar to the Eskimos. But be the origin of the name what it may, it certainly does not belong to the language of the nation, who invariably call themselves *Inu-it* (pronounced *Ee-noo-eet*), or " the people," from *i-nuk* " a man," though families or tribes have, in addition, local designations.

The Eskimos offer an interesting study to the ethnologist, on account of the very great linear extent of their country, — of their being the only uncivilised people who inhabit both the old and new continents, — and of their seclusion to the north of all other American nations, with whom they have a very limited intercourse; so that their language and customs are preserved more than any other from innovations.

They are truly a littoral people, neither wandering inland, nor crossing wide seas; yet the extent of coast-line which they exclusively possess is surprising. Commencing at the Straits of Bellisle, they occupy the entire coast of the peninsula of Labrador, down to East Main in Hudson's Bay; also, both sides of Greenland, as far north as they have been examined; and they also inhabit the islands which lie between that land and the continent, and bound Baffin's Bay and Davis's Straits on the west. On the main shore of America, they extend from Churchill, through the Welcome, to Fury and Hecla Straits; thence along the north shore to Beering's Straits, which they pass, and follow the western coast, by Cook's Sound and Tchugatz Bay, nearly to Mount St. Elias; members of the nation have also possessed themselves of the Andreanowsky Islands, Unalashka and Kadiak. They even cross the Straits of Beering, a part of the nation dwelling on the Asiatic coast, between the Anadyr and Tchukotsky Noss, where they are known to the Russians by the names of *Namollos* or *Sedentary Tchuktche*. Outside of Beering's Straits on the North Pacific, their language and customs have undergone considerable changes, as we shall have occasion to notice; but elsewhere there is no substantial va-

z 3

riation in either, the modes of life being uniform throughout, and the differences of speech among the several tribes not exceeding in amount the provincialisms of English counties.

The Greenlanders have been known to Europeans longer than any of the other North American nations, and full accounts of their manners and customs have been given to the world long ago. All the recent voyages in search of a north-west passage, also, contain characteristic portraits and descriptions of the Eskimos that reside on the west side of Davis's Straits and Melville peninsula. I shall not, therefore, attempt a systematic account of the nation, but shall confine myself chiefly to what fell under my personal notice in the central parts of the northern coast-line, where the Eskimos, from their position, have little or no intercourse with other nations, and have borrowed nothing whatever, either from the Europeans or 'Tinnè, the conterminous Indian people.

The faces of the Central Eskimos are in general broadly egg-shaped, with considerable prominence of the rounded cheeks; but few or no angular projections even in the old people. The greatest breadth of the face is just below the eyes; the forehead is generally narrow and tapers upwards; and the chin conical, but not acute; most commonly the nose is broad and depressed, but it is not always so formed. Both forehead and chin in general recede, so as to give a more curved profile than is usually to be observed in any variety of the Caucasian race, or among the male Chepewyans or Crees, though some of the female 'Tinnè have countenances approaching to the egg-shape. As contrasted with the other native American races, their eyes are remarkable, being narrow and more or less oblique. Their complexions approach more nearly to white than those of the neighbouring nations, and do not merit the designation of "red," though from exposure to weather they become dark after manhood. As the men grow old, they have more hair on the face than Red Indians, who take some pains to eradicate it, but I observed none with thick bushy beards or whiskers like those of an European who suffers them to grow. An inspection of the portraits in "Franklin's Second Overland Journey," and in "Back's Great Fish River," will show that in elderly individuals both the upper lip and chin have a tolerable show of hair, though none have the flowing beard which was productive of so much benefit to Richard Chancellor and his countrymen.

Dr. Pickering says of the Mongolian, with which, in common with other ethnologists, he classes the Eskimos and the major part of the other American nations, that both sexes have a feminine aspect;

that the stature of the men and women is nearly the same ; and that the face of the male is pre-eminently beardless. These peculiarities are but faintly developed among the Central Eskimos, and the females are uniformly conspicuously shorter than the males. Most of the men are rather under the medium English size, the defect in height being, perhaps, attributable to a disproportioned shortness of the lower extremities, though this opinion was not tested by measurements. They are broad-shouldered, and have muscular arms ; so that, when sitting in their kaiyaks, they seem to be bigger men than they do when standing erect. Some individuals, however, would be considered to be both tall and stout even among Europeans, and they certainly are not the stunted race which popular opinion supposes them to be. The comparative shortness of the females is common to them and the neighbouring 'Tinnè (Hare Indians and Dog-ribs), whose women are of small stature.

In both sexes of Eskimos the hands and feet are small and well-formed, being less than those of Europeans of similar height. The boots which we purchased on the coast were seldom large enough in the feet for our people, none of whom were tall men.

The Central Eskimos, when young, have coun-

tenances expressive of cheerfulness, good nature,
and confidence; and the females, being by no
means inclined to repress their mirth, are wont to
display a set of white teeth that an European belle
might covet. The elderly people have features
more furrowed than those we see in civilised life,
as we might expect when the passions are not
habitually repressed; and in some of the old men
the lines of the countenance denote distrust and
hatred. These ill-favoured individuals were, happily, not numerous, and several of the patriarchs we
communicated with had a truly benevolent aspect.
The weather-beaten faces of some of the old women,
gleaming with covetousness, excited by seeing in
our possession wealth beyond the previous creations
of their imagination, lead one to believe that the
poet who sang "Old age is dark and unlovely" had
drawn his picture from a people equally hard and
unsoftened by the cultivation of intellect; and I
feel no surprise that Frobisher's people should have
suspected the unfortunate elderly woman who fell
into their hands of being a witch, while they let
the young one go free.

Year after year sees these people occupied in a
uniform circle of pursuits. When the rivers open
in spring they resort to rapids and falls, to spear
the various kinds of fish that ascend the streams at
that period to spawn. At the same date, or a little

earlier in more southern localities, they hunt the rein-deer, which drop their young on the coasts and islands while the snow is only partially melted. Vast multitudes of swans, geese, and ducks, resorting to the same quarters to breed, aid in supplying the Eskimos with food during their short but busy summer of two months. In the beginning of September the rein-deer assemble in large bands and commence their march southwards; and then the Eskimos reap a rich harvest by waylaying them at established passes on the rivers or narrow places of a lake. On parts of the coast frequented by whales, the month of August is devoted to the exciting pursuit of these animals, a successful chase ensuring a comfortable winter to a whole community. Throughout the summer, the families associated by twos and threes live in tents of skins, and generally enjoy abundance of food, while they carefully lay up what they cannot consume for after use. In this respect they are more provident than the Hare Indians, or Dog-ribs, who seldom trouble themselves with storing up provisions. This difference of the habits of the two nations, which greatly influences their general characters, has perhaps originated in the different circumstances in which they are placed. The Eskimos, wintering on the coast, are in darkness at mid-winter: the rein-deer and musk-oxen have then retreated into the 'Tinnè lands, and fish cannot at that season be procured in their waters; life, therefore, can only be maintained in an Eskimo winter by stores provided in summer.

In the country of the 'Tinnè, on the contrary, the winter fishery is productive, and animals are by no means scarce at that season, but they require to be followed in their movements by the hunter and his family often to a great distance. In such a case, any surplus of food that has been procured must be placed *en cache*, as the term is, where it is exposed to the depredations of *wolverenes*, or the still more irresistible attacks of their hungry fellow-countrymen, who are wont to track up a successful hunter in order to profit by his labours. The 'Tinnè, therefore, have practically decided that it is better for them to live profusely while they have venison, and then to go in search of more. Were they to be content with the product of their fisheries, they might build villages, and live easily and well, so productive are the boundless waters of the north; but they like variety of diet, and prefer the chase, with the hazard of occasional starvation which follows in its train.

The villages of the Eskimos are, therefore, a feature in their domestic economy in which they differ wholly from their neighbours. The houses are framed strongly of drift timber, are covered thickly with earth, and are used only in winter. They have no windows, and are entered by a low side door, or, when they stand in situations where the drift-snow lies deep, by a trap-door in the roof. The floor is laid with timber, and they have no fire-places; but a stone placed in the centre serves for a support to the lamp, by which the little cooking that is required is performed. For the site of a village, a bold point of the coast is generally chosen where the water is deep enough to float a whale; and to the eastward of Cape Parry, where we saw no whales, we met with no villages, although solitary winter-houses occur here and there on that coast. The association of a number of families is necessary for the successful pursuit of the whale. When the villagers of the estuary of the Mackenzie, or of Cape Bathurst, are fortunate enough to kill one or more of these marine beasts, they revel in greasy abundance during the dark months, and the ponds and the soil around are saturated with the oil that escapes.

In March the seals have their young, and soon afterwards they become the principal objects of chase to the Eskimos, who greatly esteem their dark and unsightly flesh, reckoning it as choice food. The seal, being a warm-blooded animal, respiring air, requires a breathing-hole in the ice, which it has the power of keeping open in the severest frosts, by constant gnawing. It is a watchful creature, with acute senses of sight and hearing; but it is no match for the Eskimo hunter, who has carefully studied all its habits from his infancy. As the days lengthen, the villages are emptied of their inhabitants, who move seaward on the ice to the seal hunt. Then comes into use a marvellous system of architecture, unknown among the rest of the American nations. The fine, pure snow has by that time acquired, under the action of strong winds and hard frosts, sufficient coherence to form an admirable light building material, with which the Eskimo master-mason erects most comfortable dome-shaped houses. A circle is first traced on the smooth surface of the snow, and the slabs for raising the walls are cut from within, so as to clear a space down to the ice, which is to form the floor of the dwelling, and whose evenness was previously ascertained by probing. The slabs requisite to complete the dome, after the interior of the circle is exhausted, are cut from some neighbouring spot. Each slab is neatly fitted to its place by running a flenching-knife along the joint, when it instantly freezes to the wall, the cold atmosphere forming a most excellent cement. Crevices are plugged up, and seams accurately closed by throwing a few shovelfuls of loose snow over the fabric. Two men generally work together in raising a house, and the one who is stationed within cuts a low door, and creeps out when his task is over. The walls, being only three or four inches thick, are sufficiently translucent to admit a very agreeable light, which serves for ordinary domestic purposes; but if more be required a window is cut, and the aperture fitted with a piece of transparent ice. The proper thickness of the walls is of some importance. A few inches excludes the wind, yet keeps down the temperature so as to prevent dripping from the interior. The furniture, such as seats, tables, and sleeping-places, is also formed of snow, and a covering of folded rein-deer skin, or seal skin, renders them comfortable to the inmates. By means of antechambers and porches in form of long, low galleries, with their openings turned to leeward, warmth is insured in the interior; and social intercourse is promoted by building the houses contiguously, and cutting doors of communication between them, or by erecting covered passages. Storehouses, kitchens, and other accessory buildings, may be constructed in the same manner, and a degree of convenience gained which would be attempted in vain with a less plastic material. These houses are durable, the wind has little effect on them, and they resist the thaw until the sun acquires very considerable power.

The success of the seal-hunt depends much on

the state of the ice, and should it fail, great misery results; the spring being, in fact, the time of the year in which the Central Eskimos incur the greatest risk of famine. When the thaw lays the ground in the valleys bare, rein-deer and wild-fowl return to the sea-coast, and plenty follows in their train.

It will be evident, from the account of the yearly round of the lives of these people, that their movements are restricted to narrow limits, as compared with the 'Tinnè, who pursue the chase over tracts of country hundreds of miles in diameter, as necessity, fear, or caprice, drives them. A strict right to hunting-grounds does not seem to be maintained by the several members of the widely spread 'Tinnè nation, so as to hinder several tribes from resorting to the same districts in pursuit of deer, and meeting each other in amity, unless an actual feud exists. Thus our presence at Fort Confidence was sufficient to determine various bands of Hare Indians, Dog-ribs and Martin-lake Indians, to resort to the north-eastern arm of Great Bear Lake; and but for a deadly feud with the Dog-ribs, which twenty years ago greatly reduced the numbers of our old friends, the Copper Indians, we should have had their company also. The Eskimos, on the contrary, have a strong respect for their territorial rights, and maintain them with firmness. We learned at Cape Bathurst, that each

head of a small community had a right to the
point of land on which his winter house or cluster
of houses stood, and to the hunting grounds in its
vicinity.   We had also evidence, at various places
on the coast, of the unwillingness of these people
to appropriate the goods of their absent neighbours,
even when we, not knowing the proper owner,
tempted them by the offer of a price much beyond
the value of the article in their eyes.   The answer
on such occasions was, "That belongs to a man
who is not here."   We also saw on the coast
stages on which provisions, furs, lamps, and other
articles were placed, while the owners had gone
inland; and hoards of blubber, secured from animals
by stone walls, but without any attempt at con-
cealment.   "Tiglikpok" (he is a thief) is a term
of reproach among themselves; but they steal
without scruple from strangers, and with a dexterity which training and long practice alone can
give.   Nor did they appear ashamed when detected, or blush at our reproofs.   I believe that on
this point their code is Spartan, and that to steal
boldly and adroitly from a stranger is an act of
heroism.

In personal courage, the Eskimos are superior to
the Chepewyans, Crees, or any other Indian nation
with whom I am acquainted.   The Hare Indians
and Dog-ribs dread them, and even when much

superior in numbers, would fly on their approach. Nor do the firearms which the bolder Kutchin have lately acquired enable that people to lord it over the Eskimos, or encroach on their grounds.

The populous and turbulent bands which inhabit the estuary of the Mackenzie carry on a traffic with the Western Eskimos from the neighbourhood of Point Barrow and Beering's Straits, whom they meet midway on the coast; and though often at feud with the Kutchin have occasionally commercial relations also with them. But they who dwell to the eastward of Cape Bathurst communicate with none of their own nation except the families in their immediate vicinity, and speak of the distant Eskimos as of a bad people. The reputation of the *Kablunaht* or *Kablunèt* (white men) is superior among them to that of the remote tribes of their own nation. With the *Allani-a-wok*, as they term the inland Indians, they have no intercourse whatever.

The Central Eskimos have had no traffic with Europeans, except with those employed on the recent voyages of discovery, until the last year (1849), when a family from the coast to the west of the Mackenzie, having gone inland with a party of Kutchin, were visited at their tents by a trader sent out from La Pierre's house, which is an outpost of the Hudson's Bay Company established on the western Rat River.

Articles of Russian manufacture, procured by barter coastwise, were traced by us in an easterly direction no further than Point Atkinson.   Previous to the recent establishment of the Russian Fur Company's posts in the vicinity of Beering's Straits, the objects exchanged at Barter Island, on the 144th meridian, were brought on the Asiatic side from the fair of Ostrownoie near the Kolyma, by the Tchuktche, who passed them in the first instance to the Eskimos of Beering's Straits, by whom they were bartered at the island in question, for furs brought thither by the Eskimos of the estuary of the Mackenzie.   In like manner various wares of English make found their way, through the Kutchin and Mackenzie River Eskimos, coastwise to the Russian establishments on the Pacific.

From the predilection for commercial pursuits shown by the Eskimos, Von Bäer compares them to the Phœnicians, and, referring back to very early times, finds traces of their voyages along the eastern coasts of America, as far south as the present state of Massachusetts.   There the Scandinavian discoverers of Vinland (Rhode Island) had many skirmishes with the Skrellings (*Skrällingern*), whose identity with the Eskimos Von Bäer considers as established by the recorded descriptions of their personal appearance and dress, and the appellation given to them being the same as that applied to the Greenlanders.

From Beering's Straits, eastward as far as the Mackenzie, the males pierce the lower lip near each angle of the mouth, and fill the apertures with labrets resembling buttons, formed often of blue or green quartz and sometimes of ivory. Many of them also transfix the septum of the nose with a dentalium shell or ivory needle. These ornaments have perhaps been adopted from the Kutchin and Pacific coast tribes south of Mount St. Elias, since they have not extended to the Eskimos of Cape Bathurst or more eastern members of the nation. Most of the women are tattooed on the chin, but they have not adopted the unsightly gash and extension of the under lip on which the Kolushan ladies pride themselves.

Unlike the Hare-Indian and Dog-rib women, who neglect their personal appearance, the Eskimo females turn up and plait their hair tastefully, ornament their dresses, and evidently consider their toilet as an important concern : hence we may judge that more deference is paid to them by the men. Egede informs us, that unmarried Greenland women are modest, both in words and deeds, but that greater laxity exists among the wives, with the connivance of their husbands, who are not jealous. I fear that so much, scanty as the praise is, cannot be justly said in favour of the fair sex on the northern coast. The gestures and signs made by young and old when they came off in the *umiaks* were most indelicate, and more than once a wife was proffered by her husband without circumlocution in the presence of his companions and of the woman herself. I understood, indeed, from Augustus, our interpreter in 1826, that such an offer was considered by the nation as an act of generous hospitality; and similar customs are said to exist among the inhabitants of Tartary.

Almost all savage people are excellent mimics, and the Eskimos are not defective in this accomplishment. They imitated our speech and gestures with success and much drollery; and the men excel the other native Americans in the art of grimacing. When they wish to defy strangers who intrude into their country, they use the most extraordinary gestures and contortions of the body and limbs, making at the same time hideous faces. This was evidently practised systematically to terrify invaders; for such as resorted to it on their first interview with us, the moment that they were made to understand our friendly intentions, instantaneously relaxed their features into a broad, good-natured grin, and came alongside our boats without further hesitation.

The dress of the two sexes is much alike, the outer shirt or jacket having a pointed skirt before and behind, those of the females being merely a little longer. The Kutchin also wear these pointed skirts, but they have not been adopted by the Hare Indians or any of the Chepewyan tribes, who in common with the more southern Indians cut their shirts or frocks evenly round at the top of the thigh. I suspect that the long skirts of the Kutchin or Eskimos have given origin to the fabulous account of men with tails, thought by the Kolushes of the Pacific coast to inhabit the interior in the direction of Mackenzie's River.

The Eskimo boots are also peculiar to the nation, being made of seal-skin so closely sewed as to be water-tight, and coming up to the hips like those used by fishermen in our own land. The Chepewyans and Crees manufacture no leather that resists water; the deer-skin dressed by them like shammy absorbs water like a sponge, and hardens and spoils in drying. Neither have these Indians boots, but merely shoes or mokassins, with soft tops that wrap round the ankle, and are unconnected with the leggins or trowsers.

The Eskimos show much skill in the preparation of whale, seal, and deer-skins, using the first for thongs and lines employed in the capture of sea-beasts, also as harness for dog-sledges, soles for boots, and other purposes where strength and durability are required.

Their skin *kai-yaks* and *u-mi-aks* are also peculiar to the nation, and can be formed only by a people who dress hides so as to be waterproof. The kaiyaks are impelled by a double-bladed paddle, used with or without a central rest, and the umiaks with oars; neither of which are employed by the inland Indians, except where they have been adopted from Europeans. The use of a light waterproof outer dress, formed of the intestines of the whale, and secured to a ledge round the aperture of the kaiyak so as perfectly to exclude the water in a stormy sea, is also an Eskimo invention; and the address which is acquired in the management of the light, swift, but unstable kaiyak, contributes to the education of a race of fearless seamen.

The dogs of the Eskimos along their whole line of coast are superior in strength to those of the neighbouring nations, and are used in sledges and also in the chase of rein-deer and musk-oxen.

With respect to the religion of the Eskimos I could obtain personally no satisfactory answer to my inquiries; but it is certain that belief in witchcraft and the agency of evil spirits prevails throughout the nation, except in Greenland and Labrador, where demon worship has been combated by Christianity. Connected with this belief is the Shamanism, or influence which certain individuals claim to possess over the evil spirits. Sorcery has been reduced to a system on the shores of Beering's Sea; and that it is not unknown even on the Labrador coast, the following words, collected from an Okkak dictionary, will show. *Ange-kok,* " a shaman;" *Elihètak,* " one killed by sorcery;" *I-yèrok,* " the devil's servant or messenger;" *Nang-iner-minik,* "an appearance produced by a sorcerer;" *Torngak,* " a devil or evil spirit;" *Torngiwok,* " he performs the office of a sorcerer."

As to intelligence and susceptibility of civilisation, I consider the Eskimos as ranking above the neighbouring Indian nations, though my personal experience on this head, being confined to the interpreters employed on the several expeditions to which I have been attached, is perhaps too limited to found much upon. These individuals, however, showed a docility, industry, steadiness of purpose, a ready adoption of European customs, and an amiability which I did not observe among the Northern Indians or Crees in the course of several years' study of their characters.

The success of the Moravian missionaries, in introducing Christianity and the arts of reading and writing among the population of the Labrador coast, is a strong inducement to attempt an extension of the same system of instruction to the

A A 4

well-fed multitudes that frequent the estuary of the Mackenzie.

The origin of the Eskimos has been much discussed, as being the pivot on which the inquiry into the original peopling of America has been made to turn. The question has been fairly and ably stated by Dr. Latham, in his recent work "On the Varieties of Man," to which I must refer the reader; and I shall merely remark that the Eskimos differ more in physical aspect from their nearest neighbours, than the red races do from one another. Their lineaments have a decided resemblance to the Tartar or Chinese countenance. On the other hand, their language is admitted by philologists to be similar to the other North American tongues in its grammatical structure; so that, as Dr. Latham has forcibly stated, the dissociation of the Eskimos from the neighbouring nations, on account of their physical dissimilarity, is met by an argument for their mutual affinity, deduced from philological coincidences.

The comprehensiveness of the Eskimo language and its artificial structure are curious when we take into our consideration the isolated position of the people, and the few objects that come under their observation. In 1825, I devoted the whole winter to the formation of an Eskimo vocabulary and grammar, with the aid of our very intelligent interpreter Augustus, who was a native of the shores of Sir Thomas Roe's Welcome, and having resided at Churchill, had acquired the power of expressing his meaning in very tolerable English. The book containing the results of his labours and mine was unfortunately stolen from me in the following summer by the Eskimos of the estuary of the Mackenzie; but through the kindness of the Reverend Peter Latrobe, the philanthropic secretary of the Moravian Mission, I was provided for use on the present expedition with an excellent grammar, and a pretty full dictionary, formed by some of the industrious missionaries of the Labrador coast. By carefully perusing these volumes, together with Captain Washington's extensive vocabulary, published under sanction of the Admiralty in 1850, I feel justified in maintaining the assertion I have already made, that the Eskimo language does not materially vary throughout a line of coast longer than that which any other aboriginal people possesses. Many seeming discrepancies I have been able to trace to the genius of the language, by which the same object receives a distinct appellation for every different aspect and condition which it assumes; and the formers of the vocabularies have seldom given the precise translation such a language requires. Thus *a-niö* signifies "the snow;" *ap-ut*, "snow," a general name for snow on the ground,

whence *ap-uti-tut,* " as white as snow;" *kan-ek,*
" snow falling;" *aki-lokak,* " new fallen white
snow;" *auma-yali-wok,* " a great fall of damp
snow:" *siko,* "ice;" *tu-wak,* "solid ice;" *nilak,*
"light ice;" *ka-cho-ak,* "drift-ice;" *sir-mek,* "thin
ice." We have already remarked, that the Eskimos
of Labrador and Beering's Straits retain the name
of the musk-ox, though the Central Eskimos alone
come into contact with the animal (page 322.).*

The inhabitants of the north-western coasts from
Tchugatsky Bay (or as it is named in the English
charts, Prince William's Sound), northwards, including the peninsula of Alaska and the islands in
Beering's Sea and Straits, are considered by Baron
Wrangell, Baër, and others acquainted with them,
to be of the Eskimo stock.† Captain Beechey

* The following are some of the local designations of tribes
of the Central Eskimos. The *A-hak-nan-helet* reside near
Repulse Bay; the *Ut-ku-sik-haling-mè-ut,* or "Stone-kettle Eskimos," live further to the westward; the *Kang-or-mè-ut,* or
"White Goose Eskimos," dwell to the eastward of Cape Alexander; those who frequent the mouth of the Coppermine
River call themselves *Na-gè-uk-tor-mè-ut,* or "Deer-Horn Eskimos;" and the numerous tribe that resorts to the eastern outlet
of the Mackenzie call themselves *Kittè-gà-re-ut,* or "inhabitants
of land near the mountains."

† "The inhabitants of the Aleutian Islands (*i. e.* Beering's
and Copper Islands), of the Rat Islands, Andreanowsky Islands,
and Prebülowüni Islands, of Unalaska and Kadiak, are all Eskimo; a fact which numerous vocabularies give us full means
of ascertaining. In respect of the difference of speech between

believes that the Western Eskimos who meet the Mackenzie River tribes at Barter Island have their western boundary at Cape Barrow ; there they have commercial intercourse with the tribes described by Wrangell and Bäer, who, in their turn, barter with the Asiatic Tchuktche and with the Russians settled on the American coast, and their neighbours the Kolushans. The tribes crowded together on the shores of Beering's Sea within a comparatively small extent of coast-line exhibit a greater variety, both in personal appearance and dialect, than that which exists between the Western Eskimos and their distant countrymen in Labrador; and ethnologists have found some difficulty in classifying them properly. The appellations they have assumed, or which have been bestowed upon them correctly and incorrectly, have increased the confusion. They are, however, like the other Eskimos, a littoral people, who, in their skin kaiyaks, pursue all kinds of sea-beasts, — seals, sea-lions, walruses, polar-bears, sea-otters, and whales, — clothing themselves in their spoils and in bird-skins, and making much less use of the leather of the rein-deer skin than their southern and eastern neighbours of a different stock. The *Tchugatschih* of King William's Sound are the most southern of particular islands, there is external evidence that it is considerable."—*Dr. Latham, Varieties of Man, &c.*

several tribes, and state that, in consequence of some
domestic quarrels, they emigrated in recent times
from the Island of Kadyak*, and they claim, as
their hereditary possessions, the coast lying between
Bristol Bay and Beering's Straits.   They believe
that their nation originally sprung from a dog, in
which respect they agree with the Chepewyan
tribes, and differ from the Kolushes.

The Tchugatchih are of middle stature, slender,
but strong; with skins often brown, but in some
individuals whiter than those of Europeans, and
with black hair.   The men are handsomer than the
women.   They pierce the under lip and septum of
the nose, filling the apertures with corals, shells,
bones, and stones.   Their manners were originally
similar to those of the *Kuskutchewak* and other
communities living more to the north ; but in later
times they have carried off the women of the more
southern tribes, and from their intermarriages with
the captives, combined with their long intercourse
with the Russians, their opinions, customs, and
features have undergone a change, so that they
have now a greater resemblance to the inland
Indians than to the Northern Eskimos.

Bäer's work, which is my chief authority with
regard to the inhabitants of Russian America, contains some interesting details of the habits of the

* *Kikhtak* of the English maps.

Kuskutchewak alluded to above, from which I shall make a few extracts for the purpose of comparison with the better-known manners and customs of the Eastern Eskimos. The *Kuskutchewak* inhabit the banks of a river which falls into the sea on the 60th parallel, between the island of Nuniwak and Cape Newenham. They are neither a nomadic nor hunter folk, but dwell in winter in stationary villages built on the river, and in summer disperse themselves inland to collect provisions. They have a strong attachment to their ancestral abodes. Their winter dwellings are partly sunk in the earth, as on the Eskimo coasts, but no where else to the eastward of the Rocky Mountains. On the west coast this mode of building extends as far south as Unalaschka ; and in Cook's Third Voyage there is a representation of a winter house at that place far superior in size, accommodation, and furniture, to any that we saw on the northern shores.

In each village of the Kuskutchewak there is a public building, named the *Kashim*, in which councils are held and festivals kept, and which must be large enough to contain all the grown men of the village. It has raised platforms round the walls, and a place in the centre for the fire, with an aperture in the roof for the admission of light.*

* In Franklin's Second Overland Journals there is a plan of the Point Atkinson kashim, which answers to the above description.

I have mentioned such a building as existing at Point Atkinson (page 254.), but that was of inferior size, being indeed suited to a smaller community. In the language of the Labrador Eskimos *Kashiminwik*, or *Kashimin-wikhak*, signifies, "a place where men assemble in council;" and *Kaschim-i-ut*, "an assemblage of men for council;" from which we derive additional evidence of the national identity of the two people.

The kashim is the sleeping apartment for all the adult able-bodied males of the village, who retire to it at sunset; while the old men, women, children, and the shaman sleep in the ordinary dwellings. Early in the morning the shaman goes to the kashim with his drum, and performs some religious ceremony, varied as his fancy prompts, for the shamanism of the tribes of the Eskimo stock is said not to be guided in its ceremonials by any fixed practice.* The only women who are allowed to enter the kashim to eat with the men are those

---

* Augustus informed us that in his tribe, which occupies the coast of Hudson's Bay between Churchill and Knap's Bay, there were sixteen men and three women who were acquainted with the mysteries of shamanism. The women exhibited their skill on their own sex only. When the shaman was sent for to cure a sick person, he shut himself into a tent with his patient, and, without tasting food, sung over him for days together. The shamans also swallowed knives, fired bullets into their bodies, and practised various other deceptions to show their powers.

who have been initiated in a certain formal manner.

Feasts are held in the kashim, and particularly a great festival or harvest-home which recurs annually at the close of the autumn hunting season. Then the produce of the chase of each hunter is proclaimed before the assembly in detail, down to the small birds or mice killed by the children, and the generosity of the contributors to the feast is lauded. Many are thereby excited to give profusely, and to pinch themselves and their families for the whole of the ensuing winter. Minor feasts are held on various occasions, and the hospitality of the Kuskutchewak and neighbouring tribes is said to be very great, not only on festival occasions, but at all other times.

On the murder of a relation, retaliation is decided upon at a council held in the kashim, and is generally blood for blood. In their wars they do not slay old people or children, and instead of killing women, they lead them into slavery. On the north coast, in 1826, we observed that the old men and women were more forward in provoking a fray, in anticipation of plunder, than the young men, and perhaps they reckoned upon personal immunity in the contest. Disputes between parties of long standing are settled by dual combat in a ring of the people. Augustus, our interpreter, told me that the Eskimo of the Welcome decided their quarrels by alternate blows of the fist, each in turn presenting his head to his opponent; and Cunningham says, that the natives of New South Wales have a similar practice, but use the waddie instead of the unarmed hand, their thick skulls being able to resist blows with that formidable weapon. Both people consider it cowardly to evade a stroke. In these primitive methods of settling their points of honour we may perceive the germ of the medieval combats in lists, and of the more absurd modern duels, which the light of Christianity has not yet abolished.

When a Kuskutchewak hunter returns from the chase, he steps from his kaiyak or dog-sledge, and goes straight to the kashim, while his wife dries and secures the kaiyak, or unharnesses the dogs, and lays up the produce of his hunt. She cooks for him, and makes and mends his clothing. The husbands visit their wives, like the Spartans, by stealing out of the kashim at night, when the others have gone to rest. Every hunter preserves some remembrance of each rein-deer that he kills. He either scratches a mark on his bow, or draws out a tooth of the beast, and adds it to a girdle which he wears as an ornament.

The mode of treating infants is one of the national customs of a people that changes most slowly. It does not appear that any branch of the Eskimo nation flatten the heads or repress the growth of the feet of their children, like the Tchinuks and Kutchin. The Central Eskimo women carry their nurslings in the hoods of their shirts, and the figures in Cook's Third Voyage show that custom to be practised as low down as Unalaschka.

The Kuskuchewak are passionately fond of the vapour bath, and often use it three or four times a day, occasionally in the kashim, more frequently in small enclosures, which can be formed in every hut, and in which the steam is raised by throwing water on hot stones. If a father happens to be on bad terms with his grown-up son, he invites his most intimate friend into the bath, discloses his grievance to him, desires him to inform his son why his father is displeased, and what he ought to do to appease him. A secret which no one will tell elsewhere is revealed in the bath. This is also, I believe, a Turkish custom.

The Kuskuchewak can indicate the times of day or night with great accuracy, and they can even distinguish some stars and planets.*

Before concluding my extracts from Bäer relating to this tribe, I may remark that mammoth-teeth are numerous in crevices of the sandy banks

---

* Viz., *Tuntonok* (Rein-deer), the Great Bear; *Mi-seuschit* (the rising), Orion; *Ka-wegat* (the Fox-earth), the Pleiades; *As-guk,* Aldebaran; *Uleuch-tugal-ya* (Fox and Hare Killer), Venus; *Ag-yach-laik* (Abundance of Wild Beasts), Sirius.

of the River Kuskokwim.   The natives have a
tradition that the great animals to which the tusks
belonged came in old times from the East, but
that they were destroyed by a shaman of the
River Kwichpak.   Some of them, however, say
that the herd was merely driven into the earth,
and that it comes up in one night of the year.
Elsewhere I have alluded to the singularity of no
tusks nor fossil bones having been hitherto discovered in Rupert's Land, though they abound on
the coast of Beering's Sea.

The various small tribes or communities nearly
related to the Kuskuchewak enumerated by Baron
Wrangell are inserted in a note at the foot of the
page.*   Their name for people or men is *Tā-tchut*,
which corresponds in signification with the Eskimo
Inu-it; and among the inhabitants of the Aleutian
Archipelago, the word is modified into *Tā-gut* and
Yagut.   The similarity of this term to the national
appellation of the Lena Yakuts of Turkish stock is
worthy of notice, though it may probably be no

* Agolegmeuten, Kiyataigmeuten or Kiyaten, Mayimeuten,
Agulmeuten, Paschtoligmeuten, Tatchigmeuten, Malimeuten,
Anlygmeuten, Tschnagmeuten, Kuwichpack-meuten, are the
designations of the communities most closely allied to the
Kuskuchewak by neighbourhood and identity of manners.   The
Tchugatschen and Kadyaken, the Inkaleuchlenäten and the
Inkaliten reside at greater distances to the southward, and have
some diversities of customs.

more than a mere accidental coincidence.* The syllable *ta* in the language of the *Kutchin*, who are the inland neighbours of the Kuskuchewak, signifies water, and *Ta-kutchi* denotes "the water or ocean people." *To* (or *Ta* in composition) means " water " in the 'Tinnè or Chepewyan tongue also.

The Sedentary Tchukche, who inhabit the shores of the Gulf of Anadyr, and assume, according to Sauer, the national appellation of *Namollo*, are a tribe of Eskimos. They seem once to have possessed the coast of Asia as far westward† as the 160th parallel, traces of Eskimo dwellings having been found up to the mouth of the Kolyma. The more powerful Rein-deer Tchukche oppress and restrain them within narrow limits, and are therefore considered as the invaders of the Namollo territory.

With the mention of the Asiatic detachment of the nation, and without entering into the question

* The Turkish words *Yacubi* and *Yakutski* signify the " Sons of Jacob," and, on applying to several well informed Turks, they could recollect no other words so similar in sound to *Yakut*. A Christian community named *Yakubi* or *Yakupi* reside at the present time in Jerusalem.

† Commencing at the east coasts of Labrador and Greenland, the Namollos become the western members of the family, as Dr. Latham has noticed ; and even in respect to Europe they are *less eastern* than their American brethren. Is *Namollo* from *nuna-mullo*, " distant land" ?

of whether it ought to be considered as the remains
of the ancient trunk, or merely a decaying branch,
we close our remarks on the Eskimos.

REIN-DEER TCHUKCHE.

Before proceeding to give an account of the
Kutchin, the second of the native nations whose
lands were traversed by the expedition, I shall
introduce a brief notice of the Asiatic Rein-deer
Tchukche*, who designate themselves by the appellation of *Tchekto*, "people."   Mr. Matiuschkin
describes them as being a remarkably strong and
powerful race, resembling the Americans in their
physiognomy.   They once owned the whole country
from Beering's Straits to some distance westward
of the Kolyma, having dispossessed, according to tradition, a once numerous people, named *Omoki*, who
are now extinct, and also the *Namollos* of the
coast.†   The advance of the Cossacks and Russians
has driven them back beyond the Kolyma into

---

* The notices of this people are taken from Matiuschkin's
description of them in "Wrangell's Expedition to the Polar
Sea."   In the orthography of the name I have followed the
English translation of Wrangell's book.   The French translator
writes *Tchouktchas*, and Bäer *Tchuktschen*.   In Cook's Third
Voyage it is written *Tschutski*, and by Dr. Latham *Tshuktshi*.

† *Omoki* has an Eskimo sound: thus *oma*, "he," *okköa* or
omoköa, "they."   And as the Central Eskimos soften *k* and *g* into
l and *m*, a little etymological coaxing might produce a word like
Namollo.

the north-eastern corner of Asia*; but they resist the invaders with firmness, and maintain a greater degree of independence than any other native Siberian tribe.

Neither the Eskimos nor any other North American nation have domesticated any animal except the dog; but the Asiatic *Tchekto* are a truly nomadic people, and have tamed the rein-deer, of which they have numerous herds. The rearing of the deer, which constitutes their wealth, requires the command of a woody country and also of barren grounds or *tundras*. Commander Moore, during the winter that the "Plover" passed in Emma's Harbour, not far from Cape Tchoukotsky, purchased rein-deer from the inhabitants of a village near his anchorage, to the great benefit of his crew, and at the low rate of twelve carcasses for a ship's musket. The Tchukches are skilful traders. Those who frequent the fair of Ostrownoie bring thither furs and walrus-teeth, and receive in return tobacco, iron articles, hardware, and beads. They are accompanied by their women and children, and bring with them their arms, skin tents, and household goods, all conveyed on sledges drawn by rein-deer. The journey occupies six months, for they have to make circuitous routes in search of pasture; and

---

* For the use of the relative terms east and west see note, p. 371.

they also visit by the way Anadyrsk and Kamenoie, where inferior markets are held. After remaining eight or ten days at Ostrownoie, they commence their return ; so that their life is actually passed on the road, allowing barely time for the necessary pre- parations and for their visits to Beering's Straits. These are made in summer in baidars, or skin- boats, and in winter over the ice on sledges, with which they carry Russian wares to the Gwosden Islands in the Straits. There they are met by people from Cape Prince of Wales, with furs and walrus-teeth, collected from the dwellers in Kot- zebue Sound, and from the inhabitants of the coast still further north. The Tchukche trade with St. Lawrence Island also, and with Ukiwok, a rock of not more than three miles in circumference, but rising seven hundred and fifty-six feet above the sea. It is destitute of vegetation, and yet a body of two hundred people, from the American coast, have formed a settlement on it, at the height of one hundred and fifty feet above the water, for the purposes of trade. They inhabit caves of the rocks, and procure clothing, tobacco, and other necessaries by the sale of walrus-teeth. Sledge Island, equally small, is also inhabited by skilful traders, who are employed by the Tchukche as factors, to exchange the articles entrusted to them for furs collected on the banks of the Kwichpak, Kuskokwim, and neighbouring rivers of America.

By this channel, previous to the formation of the Russian Fur Company, wares brought from Asia were distributed over seventeen hundred miles of American coast.

The Rein-deer Tchukche practise shamanism; and though they occasionally beat their shamans to compel them to bring about some event which they desire, this treatment may be considered as evincing a belief in the powers of the sorcerer, and in times of general fear and calamity the shamans have put forth their pretensions with great effect. In 1814, an epidemic having carried off many of the Tchukche assembled at the fair of Ostrownoie, the shamans held a consultation, and decided that Kotchen, the most respected of their chiefs, must be sacrificed to appease the wrath of the spirits. Neither presents nor severe treatment could prevail on the shamans to alter their decision, and Kotchen, like another Curtius, devoted himself to the infernal gods. The love of his people, however, was such, that none could be prevailed upon to execute the sentence, until his own son, incited by the exhortations of his father, and terrified by his threatened curse, plunged a knife into his heart, and gave the body to the shamans.[*]

The tents of the Tchukche, called *namet*, have a fire in the centre with an opening for the smoke

[*] Wrangell's Polar Sea, translated by Mrs. Sabine, p. 119.

to escape, and enclose several apartments named *pologs*, or square closets of skins, stretched over laths, and so low that the inmates must remain in a crouching position. The polog is heated by a lamp, and its temperature is so high, and the air so close, as to be scarcely endurable by a person un-accustomed to breathe so impure an atmosphere.

Dr. Latham considers the Tchukche as the northern branch of the *Koriaks*, the southern branch being named *Koræki*, which is said to be an indigenous appellation*; and he reckons the Koriaks as a division of the Peninsular Mongolidæ. It is probable that on further investigation the Rein-deer Tchukche will be found to be the connecting link between the Asiatic and American Mongolidæ. In their attachment to commercial pursuits, fondness for beads, and in their bold independent character, they have a resemblance to the *Kutchin*, described in the following chapter. The similarity of the ap-pellation Tchukche, derived from *Tchekto*, " people," to *Tchutski* or *Ta-kutski*, "water-people," tribes of the Kutchin, is, however, in Dr. Latham's opinion, merely an indirect glossarial affinity. The great variety of dialects which prevail in the Aleutian Archipelago and neighbourhood of Beering's Straits is most probably the result of the active commerce there carried on, having brought several nations into contact with each other.

* *Kora*, " a rein-deer."

PLATE III.

KUTCHIN HUNTERS.

KUTCHIN HUNTERS

CHAP. XII.

ON THE KUTCHIN OR LOUCHEUX.

DESIGNATIONS. — PERSONAL APPEARANCE. — TATTOO. — EMPLOY
PIGMENTS. — DRESS. — ORNAMENTS. — BEADS. — USED AS A
MEDIUM OF EXCHANGE. — SHELLS. — WINTER DRESS. — ARMS.
— WIVES. — TREATMENT OF INFANTS. — COMPRESS THEIR
FEET. — LIVELY DISPOSITIONS. — RELIGIOUS BELIEF. — SHA-
MANISM. — ANECDOTES. — TREACHERY. — CONTESTS WITH THE
ESKIMOS. — OCCUPATIONS. — TRAFFIC. — BEADS AND SHELLS. —
TENTS. — VAPOUR BATHS. — DEER POUNDS. — ORATORY. —
TALKATIVENESS. — DANCES. — MANBOTE OR BLOOD-MONEY. —
CEREMONIES ON MEETING OTHER PEOPLE. — POPULATION OF
THE VALLEY OF THE YUKON. — SAME PEOPLE WITH CERTAIN
COAST TRIBES. — KOLUSCHES. — KENAIYERS. — UGALENTS. —
ATNAËR. — KOLTSHANEN. — PERSONS AND DRESS. — DEER-
POUNDS. — PASSION FOR GLASS BEADS. — KOLUSHES DE-
SCENDED FROM A RAVEN. — COURTSHIP. — WIVES. — REVENGE.
— MURDER. — BURN THE DEAD. — MOURNING. — DO NOT NAME
THE DECEASED. — CUSTOM CONNECTED THEREWITH. — WINTER
HABITATIONS. — JOURNEYS OF THE KENAIYER INLAND. — POR-
CUPINE QUILLS. — SLAVERY.

FROM Churchill River in Hudson's Bay round northwards to the estuary of the Mackenzie, the only
nation that the Eskimos come in contact with is
that of the 'Tinnè or Chepewyans, and even with
them they have no friendly intercourse, nor do they
meet except at the trading post of Churchill, and
within its influence. To the west of the Mackenzie,
however, another people interpose between them

and the Tinnè, and spread westward until they come into the neighbourhood of the coast tribes of Beering's Sea, which have been already noticed. On Peel's River they name themselves *Kutchin*, the final *n* being nasal and faintly pronounced. It is dropped altogether further to the westward, on the banks of the Yukon. They are the *Loucheux* of Sir Alexander Mackenzie, and the *Di-go-thi-tdinnè* of the neighbouring Hare Indians.

Of this people I have but little personal acquaintance, having had only brief interviews with the families that frequent the banks of the Mackenzie for a hundred and fifty miles or so above its delta. My information respecting them is derived from my friend Mr. Bell, who has traded with them for many years, and is the first European who penetrated into their country from the eastward; and from Mr. Murray, who is now, and has been for some seasons, resident among them. It is to this gentlemen's very able letters, which I have had the advantage of perusing, through the kindness of chief factor Murdoch M'Pherson, that I am indebted for descriptions of the tribes dwelling on the Yukon.

The few members of the nation that I saw on the Mackenzie had much resemblance in features to their neighbours the Hare Indians, but carried themselves in a more manly manner. Being, how-

PLATE IV

KUTCHIN WARRIOR & HIS WIFE.

LONDON LONGMAN & CO 1851

... regular features, ...
... ...ions than th... ...
The women resen... ...
... ... of the wife ... ...
... one that, ...
... bed them, she would
... man in any country.
... tattooed, and wh... ...
use a black pigment. ... ...
and black paints on a ... ... ceremony
... one applying th... ... his fac...
... that they may als... ... ... rema...
... a small b... ... ...
... and d to h... ...

* Mr. Isbister, speaking ... ...
River Fort, says ... ...ooking fac...
considerably above the ... ...
...
...
...
...
...
...
...

ever, merely outliers of the Kutchin, they were a less favourable example of that people than the dwellers on the Yukon that came under Mr. Murray's observation. He states that the males are of the average height of Europeans, and well-formed, with regular features, high foreheads, and lighter complexions than those of the other Red Indians.*
The women resemble the men, and Mr. Murray speaks of the wife of one of the chiefs as being so handsome that, setting aside her Indian garb and tattooed face, she would have been considered a fine woman in any country. All the females have their chins tattooed, and when they paint their faces they use a black pigment. The men employ both red and black paints on all occasions of ceremony, every one applying them according to his fancy; and that they may always have them ready, each has a small bag containing red clay and black lead suspended to his neck. Most commonly the eyes

---

* Mr. Isbister, speaking of the Kutchin who frequent Peel's River Fort, says "They are an athletic and fine-looking race, considerably above the average stature, most of them being upwards of six feet in height, and remarkably well proportioned. They have black hair, fine sparkling eyes, moderately high cheek bones, regular and well set teeth, and a fair complexion. Their countenances are handsome and pleasing, and capable of great expression. They perforate the septum of the nose, in which they insert two shells joined together, and tipped with a coloured bead at each end "—*Rep. of Brit. Ass. for* 1847, p. 122.

are encircled with black; a stripe of the same hue is drawn down the middle of the nose; and a blotch is made on the upper part of each cheek. The forehead is crossed by many narrow red stripes, and the chin is streaked alternately with red and black. The Chepewyans and Crees paint their faces in a similar manner.

The outer shirt of the Kutchin is formed of the skins of fawn rein-deer, dressed with the hair on, after the manner of the Hare Indians, Dog-ribs, and other Chepewyan tribes, but in its form it resembles the shirts of the Eskimos, being furnished with peaked skirts, though of smaller size. The men wear these peaks before and behind; the women have larger back skirts but none in front. A broad band of beads is worn across the shoulders and breast of the shirt, and the hinder part of the dress is fringed with fancy beads and small leathern tassels, wound round with dyed porcupine quills, and strung with the silvery fruit of the oleaster.* The inferior garment of both sexes is a pair of deer-skin pantaloons, the shoes being of the same piece, or sewed to them. A stripe of beads, two inches broad, strung in alternate red and white squares, runs from the ankle to the hip along the seam of the trowsers, and bands of beads encircle the ankles.

* *Elæagnus argentea.*

KUTCHA–KUTCHIN WARRIOR

The poorer sort wear only a fringe of beads, and sometimes only porcupine quills. The wealthy load themselves with beads, strung in every kind of pattern, on the breast and shoulders ; and sometimes immense rolls of this valuable article are used as necklaces. Head-bands are formed of small various-coloured beads, mixed with dentalium shells, and the same kind of shells are worn in the nose and ears. The hair is tied behind in a cue, bound round at the root with a fillet of shells and beads, and loose at the end. This cue is daubed by the tribes on the Yukon with grease and the down of geese and ducks, until, by repetitions of the process continued from infancy, it swells to an enormous thickness ; sometimes so that it nearly equals the neck in diameter, and the weight of the accumulated load of hair, dirt, and ornaments, causes the wearer to stoop forwards habitually. The tail feathers of the eagle and fishing-hawk are stuck into the hair on the back of the head, and are removed only when the owner retires to sleep, or when he wishes to wave them to and fro in a dance. Mr. Murray, when he went among these people, found that they attached nearly as much honour to the possession of their cues as the Chinese do to their pig-tails, but he in a short time acquired sufficient influence to persuade a young but powerful chief to rid himself of the cumbrous and un-

cleanly appendage; his example was followed by the
rest of his band, and will, it is to be hoped, spread
through the nation.   The mittens, which the men
always carry with them, are also adorned with
shells, and some of these expensive appendages
are even attached to their guns.   The women
wear fewer shells and beads, both of which have
a high value in the nation, especially the shells.*

In winter, shirts of hare-skin are worn, and the
deer-skin pantaloons have the fur next the skin.
On their journeys, travellers carry with them their
dress clothes, which they put on every evening
after encamping, and when they come to the
trading posts.   None of the neighbouring nations
pay so much attention to personal cleanliness and
appearance.

The arms of the men are a bow and arrow, a
knife, a dagger, and a spear, with a quiver hanging
on the left side, and suspended by an embellished
belt, which passes over the right shoulder.   Fancy
handles and fluted blades are more valued than the
good temper of a knife; and this people complain
of the trouble of sharpening a hard steel weapon.
Not so the Central Eskimos, who try one knife

* The shells, being several species of *Dentalium* and *Arenicola*,
are collected in the Archipelago lying between the Oregon and
Cape Fairweather, and pass by trade from tribe to tribe.   The
large-ribbed Dentalium is most prized.

against another, and will purchase a well tempered blade at a high price. Guns have been lately introduced among the Kutchin, and are in great demand. All the men carry powder and ball, whether they own a gun or not, and for it obtain a share of the game killed by the possessors of fire-arms. The same custom exists among the Dog-ribs.

The husbands are very jealous of their wives, but in general treat them kindly, contrasting in these respects with their neighbours. The Chepewyans treat their wives indifferently, and are jealous: the Eskimos treat them well, and are not jealous. The principal men of the Kutchin possess two or three wives each, and Mr. Murray knew one old leader who had five. Poor men, whose abilities as hunters are small, and who have been unable to accumulate beads, remain bachelors*; but a good wrestler, even though poor, can always obtain a wife. In winter the women do all the drudgery, such as collecting the fire wood, assisting the dogs in hauling the sledge, bringing in snow to melt for water, and in fact perform all the domestic duties except cooking, which is the man's office ; and the wives do not eat till the husband is satisfied. In summer the women labour little, except in drying meat or fish for its pre-

---

* Scilicet uxorem cum dote, fidemque, et amicos,
  Et genus, et formam, regina pecunia donat,
  Ac benè nummatum decorat Suadela Venusque.

servation. The men alone paddle, while the women sit as passengers; and husbands will even carry their wives to the shore in their arms, that they may not wet their feet. The Eskimo women row their own umiaks, and the Chepewyan women assist the men in paddling their canoes. On the whole, the social condition of the Kutchin women is far superior to that of the Chepewyan women, but scarcely equal to that of the Eskimo dames.

The Kutchin women do not carry their infants in their hoods or boots after the Eskimo fashion, nor do they stuff them into a bag with moss, as the Chepewyans and Crees do, but they place them in a seat of birch bark, with back and sides like those of an arm chair, and a pommel in front, resembling the peak of a Spanish saddle. This hangs at the woman's back, suspended by a strap which passes over her shoulders, and the infant is seated in it, with its back to hers, and its legs, well cased in warm boots, hanging down on each side of the pommel. The child's feet are bandaged to prevent their growing, small feet being thought handsome; and the consequence is, that short unshapely feet are characteristic of the people. A practice so closely resembling the Chinese one, though not confined, as with them, to females, may interest ethnologists.

The Kutchin live more comfortably than the

PLATE VI

KUTCHIN WOMAN AND CHILDREN

Hare Indians or Dog-ribs. They are a lively, cheerful people. Dancing and singing, in which they excel other Indians, are their favourite amusements, and they practise leaping, wrestling, and other athletic exercises. All these are called into play when different bands meet on friendly terms. They are inveterate talkers. Every new comer, as he arrives at a trading post, halts at the door of the house and makes a speech, in which he tells where he has been, what he has done, how hard he has laboured to obtain furs, and urges the propriety of his being well paid for his exertions, relating also the news he obtained from other tribes, and anything that has chequered his life or crossed his thoughts since his former visit. Established etiquette forbids any one to interrupt him until he has concluded.

Of their religious notions no full account has as yet been obtained, but they speak of good and evil spirits, and belief in shamanism is common to them, the Eskimos, and the Chepewyans. The evil spirit whose malevolence they dread is propitiated through their shamans, who profess to have the sole power of communicating with the unseen world, and of foreseeing deaths and foretelling events. Such powers clothe the shamans with authority and awfulness. Should any one have a quarrel with the members of another tribe, his death is attributed to sorcery, or, as the interpreters render it, to " evil medicine." A strong party is forthwith mustered, to seek the band which the shamans have designated, and to demand blood-money for their relative, or to avenge his death should compensation be denied. The amount claimed varies with the rank of the deceased, and the estimation in which he was held, from twenty "skins" of beads to thrice that quantity. Mr. Murray mentions a bloody instance of this superstition which occurred in 1847. A woman of the *Kutcha-kutchi* tribe dying suddenly, her death was at first attributed to the presence of white people on their lands, but the matter being debated, this opinion was overruled, and the blame was attached to a band named *Teytse-kutchi*, residing further down the river, some of whom had a dispute with the husband of the deceased. Upwards of thirty warriors started on the blood-quest, and five of the unsuspecting Teytse-kutchi happening to approach a sleeping place of the war-party were waylaid. Four of them were despatched silently on their landing, and the fifth, who was a little behind the others, not seeing his companions when he came up, suspected that evil had befallen them, and, landing on a sand-bank, interrogated the war-party across the stream. While his attention was engaged by the conversation that ensued, two of his foes carried a canoe through the willows to the other side of a point higher up the stream, and, having embarked, drifted leisurely down the river, as if they belonged to another party. On approaching the sand-bank, they called to the Teytse-kutchi man, that they were going further down, and would be glad of his company. He waited till they came up, and as he was stepping into his canoe, one of the Kutcha-kutchi tripped him up, and the other stabbed him to the heart as he lay. Having accomplished these murderous feats, the war-party resumed their voyage, but meeting afterwards only with numerous bodies of the Teytse-kutchi, they concealed their evil intentions, and returned to their own lands.

Mr. Thomas Simpson, in his "Narrative of Discoveries in the Polar Sea," relates an instance of the Peel River Kutchin demanding blood-money from the Eskimos, and receiving it for several years, for one of their countrymen, whom they asserted had died of wounds received in a contest between the two nations. The Eskimos having at length discovered that the man for whose death they had been paying was still living, reviled the Kutchin for their falsehood and extortion, and then took their revenge by killing three of the party who had come to demand the compensation for the following year.

Mr. Murray reports that the Kutchin are a treacherous people, that they never attack their enemies in open fight, and only when they consider themselves to be unquestionably superior, either by numbers or in position. They boast of their successes, but seldom tell of their reverses, which are nevertheless frequent, as their wars are chiefly with bands of their own nation, who are as wary and treacherous as themselves. By these feuds one half of the population of the banks of the Yukon has been cut off within the last twenty years. Little value is set by this turbulent people on human life, and the constant dread of ambuscade deters them from travelling except in large parties. They have not as yet imbrued their hands in European blood. Messrs. Dease, Bell, and others of the Company's officers, who have resided at Fort Good Hope and on Peel River, have used their influence, and distributed large presents among the tribes, for the purpose of establishing peace, but only with temporary success. The pretensions and arts of the shamans are fertile sources of mischief.

The Peel River Kutchin, in speaking to Mr. Bell of their contests with the Eskimos, always charged the latter with treachery, but it is more likely that they were themselves the aggressors. One of their encounters with that people deserves to be mentioned here, because of its resemblance in some particulars to the meeting of Joab and Abner recorded in the second book of Samuel. A party of each of the two nations having met, the young men rose up to dance, as if the meeting had been entirely amicable; but the Eskimos having, as they are accustomed to do, concealed their long flenching knives in the sleeves of their deer-skin shirts, drew them in one of the evolutions of the dance, and thrust them into their opponents. A general conflict ensued, in which the Kutchin were the victors, owing to their guns — that is, according to their report of the affair; but had the Eskimos been the tellers of the story, the circumstances might have been related differently.

Another incident, which occurred on the banks of the Yukon in 1845, gives us a further insight into the suspicious and timorous lives of these people. One night four strangers, from the lower part of the river, arrived at the tent of an old man who was sick, and who had with him only two sons, one of them a mere boy. The new-comers entered in a friendly manner, and, when the hour of repose came, lay down; but the sons perceiving that their guests did not sleep, and suspecting from their conduct that they meditated evil, feigned a desire of visiting their moose-deer snares. They intimated their purpose aloud to their father, and went out, taking with them their bows and arrows.

Instead, however, of continuing their way into the wood, they stole back quietly to the tent, and listening on the outside, discovered as they fancied, from the conversation of the strangers, that their father's life was in danger. Knowing the exact position of the inmates, they thereupon shot their arrows through the skin covering and killed two of the strange Indians; the other two, in endeavouring to make their escape by the door, shared the fate of their companions. This is spoken of in the tribe as an exceedingly brave action. The golden age of innocence and security is not to be sought for among a savage people ignorant of the precepts of divine truth.

The Yukon Kutchin pass the summer in drying white-fish (*Coregonus*) for winter use. For the purpose of taking these fish, they construct weirs by planting stakes across the smaller rivers and narrow parts of lakes, leaving openings in which they place wicker baskets to intercept the fish. This practice is common in Oregon and New Caledonia*, but does not exist to the eastward of the Rocky Mountains. On the other hand, the inhabitants of the Yukon are unacquainted with nets, so largely employed by the Chepewyans and Crees. The Kutchin take the moose-deer in snares, and,

* Cook observed fishing weirs at Nootka Sound. — *Third Voy.*, vol. ii. p. 281.

PLATE VII

W. H. MURRAY DEL.   M & N. HANHART LITH.

SAVĨAH, CHIEF OF THE KUTCHA-KUTCHIN

LONDON LONGMAN & CO 1851

PLATE III

W. H. MURRAY DEL.ᵗ

M. & N. HANHART LITH.

SAVĪAH, CHIEF OF THE KUTCHA-KUTCHIN

LONDON LONGMAN & Cᵒ 1851

Beads ... the price of ...

Their accounts of the ...

messagls leads to ...
standard bead, and ...
large one of white ...
in Italy only, and can ...
... blue and red ones of ...
... a small white one, ...
for ... their ...

Deerskins and ...
from the west coast in ...
... None of the Ch...

c c 2

SAV AND CHIEF OF THE NORTHE...

towards the spring, most of the nation resort to the mountains, to hunt rein-deer and lay in a stock of dried venison.

Beads are the riches of the Kutchin, and also the medium of exchange throughout the country lying between the Mackenzie and the west coast, other articles being valued by the number of strings of beads they can procure. No such near approach to money has been invented by the nations residing to the eastward of the Rocky Mountains, though their intercourse with the fur-traders has given them a standard of value in the beaver's skin. Their accounts at the posts being reckoned by the number of " beavers " they owe, and the Company's tariff fixes the value of a " beaver." To be accounted a chief among the Kutchin, a man must possess beads to the amount of 200 beavers. The standard bead, and the one of most value, is a large one of white enamel which is manufactured in Italy only, and can with difficulty be procured from thence in sufficient quantity. Fancy beads, *i.e.*, blue and red ones of various sizes, and the common small white ones, are, however, in request, for ornamenting their dresses.

Dentalium and *Arenicola* shells are transmitted from the west coast in traffic, and are greatly valued. None of the Chepewyan tribes wear nose-ornaments, neither have the latter people the same passion for beads. A supply of them is indeed sent to all the trading posts frequented by the 'Tinnè, but they are mostly purchased by the wives of Canadian voyagers or half-breeds residing in the establishments, and if desired by the natives for the same purposes they are given to them as presents, or exchanged for articles of small value, and never, I believe, for furs. The Kutchin, on the contrary, will not part with their furs unless they receive most of the price in beads or shells, as they have not yet learnt to value English cloth and blankets above their skin dresses. Ammunition, by the instructions of the Hudson's Bay Company, is given in exchange for provisions, or, when the natives are in want, gratuitously, that they may be able to support themselves.

Each family possesses a deer-skin tent or lodge, — the skins used in winter being prepared without removing the hair, that the cold air may be more effectually excluded. In summer, when the family is travelling in quest of game, the tent is rarely erected. A winter encampment is made usually in a grove of spruce firs. The ground being cleared of snow, the lodge-skins are extended over flexible willow poles, which take a semicircular form, and are transported with them from place to place. The hemispherical shape of their lodges is not altogether unknown among the Chepewyans and Crees, being that generally adopted for vapour baths, which are framed of willow poles stuck into the ground at each end; but the lodges used by these nations for dwelling places are cones, formed by stiff poles meeting at the top. The lodges of the Kutchin resemble the Eskimo snow huts in shape, and also the *yourts* of the Asiatic Anadyrski. When the Kutchin winter-lodge is raised, snow is packed on the outside to half its height, and it is lined equally high within with the young spray of the spruce fir, that the bodies of the inmates may not rest against the cold wall. The doorway is filled up by a double fold of skin, and the apartment has the closeness and warmth, but not the elegance, of a snow house. Mr. Murray remarks, that though only a very small fire is usually kept in the centre of the lodge, the warmth is as great as in a log house. The provisions are stored on the outside, under a covering of fir branches and snow, and further protected from the depredations of the dogs by the sledges being placed on the top.

Mr. Bell informed me that, on the open hilly downs frequented by rein-deer, the Kutchin have formed pounds, towards which the animals are conducted by two rows of stakes or trunks of trees extending for miles. These rows converge, and as the space between them narrows, they are converted into a regular fence by the addition of

strong horizontal bars. The extremity of the
avenue is closed by stakes set firmly in the ground,
with their sharp points sloping towards the entrance,
so that when the deer are urged vehemently forwards they may impale themselves thereon. The
hunters, spreading over the country, drive the
deer within the jaws of the pound; and the women
and children, ensconced behind the fence, wound
all that they can with arrows and spears. These
structures are erected with great labour, as the
timber has to be brought into the open country
from a considerable distance. Some of the pounds
visited by Mr. Bell appeared to him, from the condition of the wood, to be more than a century old.
They are hereditary possessions of the families by
whom they were constructed.

Mr. Murray's letters describe the meetings of
several tribes which he witnessed. On one occasion two parties who had been at war with each
other, and had not yet arranged their differences,
met at his encampment. Very long harangues
were made by different members of the two bodies
before they landed. After this, one party, stepping
from their canoes, formed a circle and pranced
round, yelling and shouting furiously. The other
party landed a little way off, and ranging themselves in Indian file, the chief in front and the
women and children in the rear, danced forward

slowly, until they came up to the others; when the whole, joining in one circle, capered for half an hour, uttering the most horrible cries, the two chiefs meanwhile keeping in the centre.

The formal dance is always in a circle, but the gestures and the songs which accompany them vary. After a ball kept up vigorously by man, woman, and child, for hours, the parties retire to their tents, and raise the song at intervals till the morning. At the festivals held on the meeting of friendly tribes, leaping and wrestling are practised.

Several other illustrations of the superstitions and manners of this people might be quoted from Mr. Murray's letters, but, as they relate only to peculiarities already mentioned, I shall restrict my extracts to two other anecdotes. A young chief was at Mr. Murray's encampment when a party of another tribe, named the *Vanta-kootchi*, arrived, one of whom had married the chief's sister, and was reported to have killed her. The young chief demanded an expiatory offering in beads for his sister's death, which was refused; and an altercation ensuing, something was said which insulted him. He immediately drew his knife and walked boldly up to the others, who would have cut him in pieces but for the immediate intervention of the white men. A few words of explanation from a hunter in Mr. Murray's employment calmed the storm. The woman had not been killed, but was drowned in crossing a river, through the upsetting of a canoe. A present of a large Eskimo spear, valued at ten skins, was made to the brother, and peace was restored.

A body of the *Han-kutchi*, residing at the sources of the Yukon, came to visit Mr. Murray early in August. Rumours of their hostile intentions preceded them. The sudden death of their chief in summer had been attributed by them to the shamanism of the *Kutcha-kutchi*, and also to the presence of white people in the country. The Canadian voyagers looked on, therefore, with apprehension, which was not quieted by the first movements of their visitors. Twenty canoes first appeared, gliding stealthily down the river to a point above the encampment, on which the party landed and assembled in silence. Mr. Murray walked up to them, and expressed his pleasure at the meeting; but, rushing past him, the whole body ran in full career to the lower end of the encampment and back again to their landing-place, shouting and whooping in a peculiar manner. Then they formed a half circle, and danced with great energy for a few minutes, beating time to their songs with their feet. Their dresses, tinkling with beads and brass trinkets, and their long

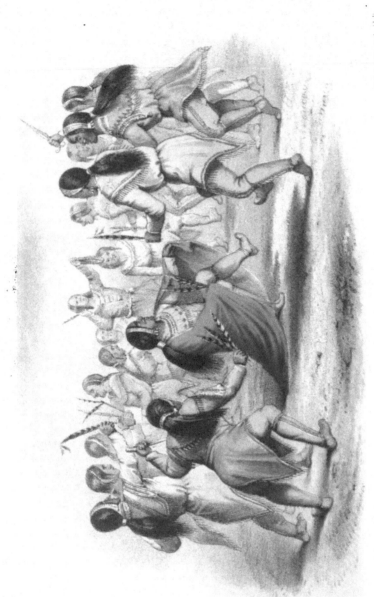

...ed ...
... ...
p... ...
... ...
the ... ...
p...ted ...
... ...
which ... ...
... ...
... ...
... ...
... and ...
f... ...
... Russians had used ... ...
... come more civil since they had ... ...
their outposts. At ... ...
the ... ...
but reti...d ... ...
...ons.

Mr. ... ...
natives, estimated the popula...
Yukon at about ... thousand
... The ... ...
the upper ... ...
the Pacific, ...
Alexander, or ... ...
... 100 n ... ...
... of about the ... ...

clouted hair shaking in the wind, gave them a wild and savage aspect. They were armed with pikes of their own manufacture, also with similar weapons, mounted with sheet-iron, procured from the Russians. On the dance ceasing, Mr. Murray presented the brother of the deceased chief with twelve inches of tobacco to smoke over the grave, which produced a favourable impression, and called forth the remark that *now* he could consider the white people as his friends. These Indians afterwards became troublesome, asked for goods on credit, and on being denied threw out some significant innuendos, saying, among other things, that the Russians had used them so at first, but had become more civil since they had cut off one of their outposts. At the great dance in the evening, the deceased chief's brother did not join the circle, but retired to a corner, and made piteous lamentations.

Mr. Murray, from information collected from the natives, estimates the population of the banks of the Yukon at about one thousand men and boys able to hunt. They are distributed as follows. Between the upper branches of the river and the coast of the Pacific, on or near the 62d parallel, reside the *Artez-kutchi,* or "tough and hard people," numbering 100 men. The *Tchŭ-kutchi,* "people of the water," of about the same numbers, inhabit the banks of Deep River, a western affluent of the Yukon, opposite Comptroller's Bay. The *Tathzey-kutchi*, "people of the ramparts," known to the traders and Canadian voyagers by the name of "*Gens du Fou*," number about 230 men, who are divided into four bands, the uppermost of which is called *Trātzè-kutchi*, "people of the fork of the river." * The *Tathzey-kutchi* inhabit a wide country, which extends from the sources of the Porcupine and Peel to those of the River of the Mountain-Men. They visit the Russians on the coast of the Pacific, and trade with the intervening tribes. The Indians of the lowlands at the influx of the Porcupine River, named *Kutcha-kutchi*, number 90 men. Further down the Yukon are the *Zēkā-thaka*, or *Zi-unka-kutchi*, "people on this side," or "middle people," numbering only 20 men. West of these reside the "people of the bluffs," *Tanna-kutchi*, 100 strong. And further down, at the influx of Russian River, are the *Teytsè-kutchi*, "people of the shade" or "shelter," of about 100 men.

* One of *'Dtchè-tà-ùt 'tinnè*, or Mountain Indians, who inhabit the conterminous mountainous country that lies between the *Nok'hannè* and the *Becàtess* ("Gull" or "Gravel River" of the voyagers), called the *Gens du Fou*, *Ey-unè 'tinnè*; and a Mountain Indian hunter, employed by Mr. Bell, called them *Tratzà-ut 'tinnè*, which is evidently another mode of pronouncing the designation they give themselves, substituting for *Kutchi* the synonymous appellation *'Tinnè*.

Nearer the mouth of the Yukon are two bands, usually called *Tlagga-silla*, or "little dogs," but not by themselves. These trade their furs with the *Kuwichpack-meuten*, mentioned in a preceding page.*

The banks of the Porcupine and country on the north of it belong to the *Vanta-kutchi*, " people of the lakes," having 80 men ; and to another band, named *Neyetsè-kutchi*, " people of the open country," who have 40 men.

Though the short Kutchin vocabulary formed by Mr. Murray, which will be printed in the Appendix, contains some Eskimo words, the language, as far as one can judge from so brief an example, is more nearly related to the 'Tinnè. How far the Kutchin are to be considered as actually a distinct nation from the 'Tinnè or Chepewyans, must be decided hereafter when their two languages are better known.

At my suggestion Mr. Murray made, by the aid of his Athabascan interpreter, the following collection of words, having a similar sound and signification in the Kutchin and Dog-rib languages.

English.	*Kutchin.*	*Dog-rib.*
One	Tech-lagga	Inch-lagga or Ingeh-lagga.
Two	Näk-heiy	Nak-hè.
A hare	Ké	Kä.

---

* Page 370.

English.	Kutchin.	Dog-rib.
Fire	Kon or Khon	Khu or Kun.
Wood	Tutshun	Tutshin.
A blanket	Tsattè	Tsat-hè.
Rein-deer (male)	Bat-zey-tcho	Bet-sich-tcho.
A dog	Chli-en or Chlyn	Chli or Thling.
My country	Sun-nun	Sa-nun-nā.
Moss	Ni-en	Ni.
A goose	Chrè	Chra.
A snow goose	Ku-kè	Ko-ka.
Thunder	Nach-thun	Nach-thun.
Willow	Khy-i	Khy-i.
Snow shoes	Ai-i or Ay-i	Ah.
A poplar tree	Tho	Tóë or T'höe.
An Indian cap	Tsè	Tsa.
White	Ta-kynè	Tek-kynè.
Man	'Tingi	'Dinnè, or 'Tennèh, or Dunneh.
A quill or feather	Teh or Tay	Tah.
A sinew	Tchè	Tlè.
You	Nuhn or Nunn	Nunney.

In this vocabulary the *ch*, except when immediately preceded by *t*, is pronounced as in the Scottish "*loch*," or Irish "*och;*" *u* is sounded as *oo* in good, except before double consonants; and *i* as *ee* in "see," or *e* in "me."

Mr. Murray remarks that, though the above words and a few names of trading goods are similar in sound, the languages of the two nations are very different. More resemblances, he thinks, might be traced through the Mountain Indian speech (*Naha 'tdinnè* or *Dtchè-ta-ut 'tinnè*) than directly between the Kutchin and Dog-rib tongues. The *Han-Kutchi*, of the sources of the Yukon, speak a dialect of the *Kutcha-Kutchi* language, yet they understand and are readily understood by the Indians of Frances Lake and the banks of the Pelly. Now these converse freely with the *Naha-* or *Dtchè-ta-ut 'tinnè*, and other Rocky Mountain tribes, whose language resembles the Dogrib tongue, and who are, in fact, acknowledged members of the Chepewyan nation. Again, the Frances Lake Indians understand the *Netsilley*, or Wild Nation, who trade at Fort Halkett, on the River of the Mountains; these again are understood by the *Sikānis;* and the *Sikānis* by the Beaver Indians, whose dialect varies little from that of the Athabascans, the longest-known member of the 'Tinnè nation.

From the great resemblance in manners, customs, and person of the *'Tnaina** or *Kenaiyer* of

* *Tnai* signifies "men," and is used when the Atnäer speak of themselves. Their Eskimo neighbours the *Kadyakers* call them *Kenai-yut*, and the Russians have adopted this latter appellation. *Koltshanen* means "strangers" in the Atnäer dialect, and *Gultzanen* "guests" in the language of the Kenaiyer. These people, coming from the interior, about the sources of the Copper River and the water-shed between it and the valley of the Yukon, have commercial relations with the Kutchin who dwell on Deep River, an affluent of the Yukon. Their accounts of the pointed skirts of the Kutchin shirts have given rise to the fable already alluded to of men with tails dwelling beyond the mountains, current among the Kenaiyer. And their reports of the canabalism of the tribes of the interior have a similar foundation. Lynn's canal is not mentioned as the ascertained

Cook's Inlet, the *Ugalakmutsi* or *Ugalents* of King
William's Sound, the *Atnäer* of the mouth of the
Copper River, the *Koltshanen* or *Galtzanen* of the
sources of that stream, and the other Kolusch
tribes as far as Tchilkat or Lynn's Canal, on the
54th parallel and 135th meridian, to those of the
Kutchin, I am inclined to consider them as all of
the same stock. Captain Cook, from whom we
have the earliest accounts of these people, remarks
their dissimilarity in person and language to the
Wakash nation who inhabit Vancouver's Island.

The *Kenaiyer* and *Ugalents* are described by Captain Cook * and Baron Wrangell as a moderate-sized

southern limit of the tribes akin to the Kutchin, but because
down to that inlet the language of the Yukon Kutchin seems to
be readily comprehended. The travelling merchants who go
from thence to the banks of the Pelly and sources of the Yukon
meet there with the *Tratze-kutchi* or *Gens du Fou*, by whom
they are understood, — as has been mentioned in page 167 of
the narrative. Dr. Scouler and others maintain, and probably
with justice, that the Kolusch language is spoken as low as
Queen Charlotte's Island and Observatory Inlet.

* " The natives who came to visit us in the Sound were generally not above the common height, though many were under
it. They were square, or strong chested; and the most disproportioned part of their body seemed to be their heads, which
were very large; with thick short necks; and large, broad or
spreading faces, which upon the whole were flat. Their eyes,
though not small, scarcely bore a proportion to the size of their
faces; and their noses had full round points hooked or turned up
at the tip. Their teeth were broad, white, equal in size, and evenly
set. Their hair was black, thick, straight, and strong; and their

race, occasionally tall, with women of equal height, and when young handsome; particulars in which they agree with the Kutchin, whose women have finer persons than the 'Tinnè females. Both these authors, however, remark that there are many deformed individuals among the coast tribes, contrary to what occurs among the inland Indians. Both sexes of the Kenaiyer and of the allied tribes powder their hair with the down of birds, and smear their faces with black, blue, and red pigments. The blue pigment was noticed by Cook, and Mr. Bell informed me that some of the Peel

beards in general thin or wanting; but the hairs about the lips of those who have them, were stiff or bristly, and frequently of a brown colour. And several of the elderly men had even large and thick but straight beards." "Their countenance indicates a considerable share of vivacity, good nature, and frankness. Some of the women have agreeable faces."— *Cook, Third Voy.*, vol. ii. p. 366.

Wrangell says, "The Kenaiyer are in general of middle size, slightly built, and betray a true American descent in their features and colour of the skin. Many among them are of gigantic stature, and I have never seen so many deformed persons among any people in the colonies."

These quotations will give the impressions the coast tribes made on two accurate observers at distant periods. With respect to the shortness and thickness of the neck, I may mention that the skeletons of the races which flatten their foreheads artificially have very short necks, the bodies of the vertebræ being unusually thin. This is doubtless the result of the process resorted to in infancy; and the same effect may be produced by masses of dirt and hair adding weight to the head.

River Kutchin possess it, but he did not ascertain
where they procured it, and I have not been able
to obtain a specimen of it, so as to ascertain its
nature.   The men wear dentalium shells in the
nose, and also ear-pendants and the unsightly
labrets.   They are a cheerful people, sing during
labour, and when it is over recreate themselves
with dancing.   Cook notices the thick short necks
of the people of Prince William's Sound.   The
muscles of the necks of the Kutchin are called
strongly into action for the support of their
weighty ties of hair, and, in consequence thereof,
increase in size ; and we might have been inclined
to attribute the disproportion noticed by Cook
to a similar cause, had he not mentioned that the
men wear their hair cropped round the neck.
Some of the women, however, clubbed it behind ;
and this looks as if, at that period, the tribe had
found out the inconvenience of the unwieldy and
uncleanly cues, and that a few only of the women
retained them.

Among the Atnäers and neighbouring tribes who
hunt rein-deer, pounds, formed of hedges converging
thus >, are in use.   Weirs and wicker-baskets for
taking fish are also constructed by the coast tribes.
The shirts of the Ugalents reach to the knee or
lower, and are cut evenly round without peaks.
They are made of skins, with the fur turned out-

wards; but the coast people wear also a waterproof outer dress, made of whale-gut or other thin membranous substance, which they have most probably adopted from their Eskimo neighbours. The dwellers of the interior do not use it, and indeed have not the material. The Koltshanen construct birch-bark canoes; but on the coast skin boats or baidars, like the Eskimo kaiyaks and umiaks, are employed. Another opening has, however, been added to the kaiyak, which is therefore longer, and carries two sitters. It would seem that the constant commercial intercourse between the coast tribes had led them to adopt whatever they thought worthy of imitation from one another.

The passion for glass beads extends from the coast to the Mackenzie. The Atnäers bury their accumulations of this treasure in the earth, and leave the hoard to their children. In Cook's time a large light-blue bead was in the greatest request.

All the coast tribes now under consideration claim descent from the Raven; and it would strengthen the opinion we entertain of their affinity with the Kutchin, if the same belief existed among the latter. This, however, has not been ascertained. The west coast Eskimo and the Chepewyan tribes agree in tracing the origin of their respective nations to the Dog. The Raven, say the Northern Kolushes, Prometheus-like, stole the elements one after another, out of which he made the world. The Kenaiyer tradition is, that the Raven also made two women, one of whom is the mother of six races, and the other of five.* It was the custom that the men of one stock should choose their wives from another, and the offspring belonged to the race of the mother. This custom has fallen into disuse, and marriages in the same tribe occur; but the old people say that mortality among the Kenaiyer has arisen from the neglect of the ancient usage. A man's nearest heirs in this tribe are his sister's children, little going to his sons, because they received in their father's lifetime food and clothing.

Courtship is a simple affair with these people. Early in the morning the lover makes his appearance at the abode of the father of the object of his choice, and, without a word of explanation, begins

* The stock descended from one female ancestor are, 1. the *Kachgiya*, from *Gekaihze* "the raven;" 2. *Tlachtana*, so called from their being weavers of grass-mats; 3. *Montochtana*, who take their name from a corner in the back part of their huts; 4. *Tschichgi*, named from a colour; 5. *Nuchschi*, " descendants from heaven;" 6. *Kali*, "fishermen;"

The races sprung from the other woman are, 1. *Tultschina*, "bathers in cold water;" 2. *Katluchtna*, "lovers of glass beads;" 3. *Schischlachtana*, " deceivers like the raven, who is the primary instructor of man;" 4. *Nutschichgi;* and 5. *Zaltana*, named from a mountain on the borders of Lake Skiläch towards the sources of the river Katnu.

to heat the bath room, to bring in water, and to prepare food. Then he is asked who he is, and why he performs these offices. In reply he expresses his wish to have the daughter for a wife, and if his suit be not rejected, he remains as a servant in the house a whole year. At the end of that time he receives a reward for his services from the father, and takes home his bride. No marriage ceremony takes place. Rich men have three or four wives. The wife, though the most industrious worker in the family, is not the slave of her husband. She may return, if dissatisfied with the treatment she receives, to her father's house, and then she takes with her the dowery the husband received at the conclusion of his year of service. The wife retains as her own property whatever she gains by her labours, and it often happens that the husband makes purchases from her. If there be several wives in a family, each has her own household stuff, which may not be meddled with by the other wives, or by any member of the household.

Banquets, accompanied by dances, songs, and distribution of presents, take place on various occasions. A man, on recovering from sickness, will give a feast for the benefit of those who have shown him most sympathy during his illness. One who spends freely on these occasions is looked up to by his fellow-countrymen, and his advice sought. This is the origin of *Toyonhood*, or chieftainship. Though the power of the petty chief does not depend on descent, it frequently passes to his heir; but submission to him is conditional, and any one may attach himself to another leader.

If a man be murdered or injured by one of his own clan, the nearest relative revenges it without seeking aid; but if the injury be perpetrated by one of another clan, the allied families are called together to consult on the defence of their honour. The feud that ensues is sometimes bloody, but seldom of long duration, and any prisoners that are taken are set free for a ransom, or retained as slaves. Before the Russians came, all the tribes were at war with the Kadyakers and others of the Eskimo nation, who on that account received from the Kenaiyer the denomination of *Ultsehna*, or *Ultsehaga*, "slaves."

One who dies is mourned by his whole clan. The mourners assemble in the dwelling of the nearest kinsman, sit round the fire, and howl. The master of the house, dressed in his best garments, leads the lamentation, having his face blackened, an eagle's feather in his nose, and a cap of eagle feathers on his head. Ringing a bell which he holds in each hand, he raises the voice of mourning, making at the same time violent contortions of the body, and stamping continually on the ground with his feet. He recounts in his song the famous actions of the deceased; stanzas are improvised by the other mourners, and sung to the accompaniment of drums. A general cry of grief is raised at the end of every verse, and during its continuance the chief mourner pauses from his exertions, and lets his head sink on his breast.

The clothing and rest of the property of the deceased are divided among mourning relatives. The body is then burnt, and the bones are collected and interred by friends who are not of kin to the deceased. At the end of the year the nearest relative celebrates a festival to the memory of the departed. From that time the dead man's name must never be pronounced in the presence of that relative, and *he* even changes the name by which the deceased had been accustomed to address him. If a relation transgresses this law, he is reproved; but if it be a more distant friend he is challenged by the kindred, and must buy himself off. Poor men will sometimes endeavour to entrap a rich relative into a breach of this custom, to obtain the redemption money. The Kenaiyer suppose that after death a man leads, in the interior of the earth, where a sort of twilight reigns, a similar life to his former one, but that he sleeps when those

on the surface are awake, and wakes when they
sleep.

One other way, among several, in which a poor
Kenaiyer endeavours to improve his condition is,
to invite his rich friends of another family to a
festival. Melted snow only is set before them, and
the relatives, watching for any sneering expressions
respecting this Barmecide feast, report them to the
host. He then rushes out with angry gestures,
and challenges his mockers, wounding himself at
the same time with an arrow, to signify that he
prefers death to degradation. This scene was expected, and the mockers, having their presents prepared, declare their readiness to make reparation.

The winter huts of the Northern Kolushes are
high, large, and roomy, built of wood, with the
hearth in the middle, and the sides divided into as
many compartments as there are families living
under the roof; the number varying from two to
six. Two or three bath rooms are constructed at
the end, and in them much of the winter is spent.
These vapour baths resemble the den of a bear,
and open into the hut by a small aperture through
which a man can creep with difficulty. They are
covered on the outside with earth, and heated
within with hot stones.

Baron Wrangell, from whom most of the preceding details have been translated, is of opinion

that the Kenaiyer came originally from the interior, bringing with them their bark canoes, and borrowing the skin kaiyaks from the Eskimo Tchugatschih, who are more expert and bolder navigators of the sea. The following notice of the journeys of the Kenaiyer, given by the author above named, will show that they must come occasionally into contact with the dwellers of the Yukon Valley, as they approach its water-shed. The people on the south side of Cook's Inlet hunt mountain sheep in the neighbouring hills. Those that dwell on the north side travel much further. Striking off to the north-east, after seven days of rapid march, or ten of ordinary travelling, in which they can accomplish about one hundred and thirty or one hundred and forty miles, they arrive at the foot of a high mountain chain, where, on the banks of Lake Knitiben, the women, children, and less skilful hunters are left. The others cross the mountains in a southerly direction, and in seven days more arrive at Lake Chtuben, on an elevated plateau not far from the source of the Suschitna, and fourteen days' march from the northern arm of Cook's Inlet.

There they kill rein-deer, which winter in that district in numbers. The Atnäer and Galzanen of Nutatlgat on the Copper River come, after ten days of rapid travelling, to the same lake. From thence the Kenaiyer go six days' march further

to a small lake, where they are met for the purposes of trade by the more distant Koltshanen,
who supply them with articles of English manufacture.   Porcupine quills, coloured by the Atnäer
with moose berries, used for embroidering seal-
skin shirts, are valued articles of commerce; and
in their traffic all parties are wary and skilful.

The fatigues of this excursion render the Kenaiyer hunters lean and exhausted; but they kill
beavers as they return home, and continue to do
so up to the beginning of winter, when they hold
the annual festival, announce the produce of the
chase, and give themselves up to recreation and
pleasure.

Slavery, to a certain extent, prevails among the
coast tribes, and the slaves, who are originally
prisoners taken in battle and not redeemed, are
transferred from one tribe to another by barter.
Captain Cook mentions, that on his first intercourse
with them none would come into his ship, until a
seaman had gone into one of their boats, when an
Indian was sent up into the ship.   This was considered by the English officers as an exchange of
hostages, but perhaps the Indians reckoned it to
be a barter of slaves, as they would not part with
the seaman until muskets were presented at them;
and when he returned on board they took their
own man and departed.   The slaves are named

Kalgen, and are occasionally sacrificed to the manes of deceased chiefs by the Koltshanen and Koluschen, but not by the Atnäer. Previous to the arrival of Europeans, the Atnäer were workers in copper, and supplied the neighbouring tribes with weapons. They have since acquired the art of fabricating Russian iron into various articles. They are a milder tribe than the other Kolushes, and are generally on good terms with the rest.

END OF THE FIRST VOLUME.

An ALPHABETICAL CATALOGUE of
NEW WORKS
in
GENERAL & MISCELLANEOUS LITERATURE,

PUBLISHED BY

Messrs. LONGMAN, BROWN, GREEN, AND LONGMANS,
PATERNOSTER ROW, LONDON.

Classified Index.

Alphabetical Catalogue

NEW WORKS AND NEW EDITIONS,

PUBLISHED BY

MESSRS. LONGMAN, BROWN, GREEN, AND LONGMANS,

PATERNOSTER ROW, LONDON.

~~~~~~~~~~~~~~~

MISS ACTON'S MODERN COOKERY-BOOK.

Modern Cookery in all its Branches, reduced to a System of Easy Practice. For the use of Private Families. In a Series of Receipts, all of which have been strictly tested, and are given with the most minute exactness. By ELIZA ACTON. New Edition: with Directions for Carving, and other Additions. Fcp. 8vo. with Plates and Woodcuts, 7s. 6d. cloth.

AIKIN.—SELECT WORKS OF THE BRITISH POETS,

From Ben Jonson to Beattie. With Biographical and Critical Prefaces by Dr. AIKIN. New Edition, with Supplement, by LUCY AIKIN; consisting of additional Selections from more recent Poets. 8vo. 18s. cloth.

ALLEN ON THE ROYAL PREROGATIVE.—AN INQUIRY

into the RISE and GROWTH of the ROYAL PREROGATIVE. By the late JOHN ALLEN, Master of Dulwich College. New Edition, with the Author's last Corrections: preceded by M. Bérenger's *Rapport* on the Work read before the Institute of France; an Article on the same from the EDINBURGH REVIEW; and a Biographical Notice of the Author. To which is added, An Inquiry into the Life and Character of King Eadwig, from the Author's MS. 8vo. 12s. cloth.

THE ARTISAN CLUB.—A TREATISE ON THE STEAM

ENGINE, in its Application to Mines, Mills, Steam Navigation, and Railways. By the Artisan Club. Edited by JOHN BOURNE, C.E. New Edition. With 30 Steel Plates and 349 Wood Engravings. 4to. 27s. cloth.

JOANNA BAILLIE'S DRAMATIC AND POETICAL WORKS.

Now first collected; complete in One Volume; and comprising the Plays of the Passions, Miscellaneous Dramas, Metrical Legends, Fugitive Pieces (including several now first published), and Ahalya Baee. Uniform with the New Edition of *James Montgomery's Poetical Works*; with Portrait engraved in line by H. Robinson, and Vignette. Square crown 8vo. 21s. cloth; or 42s. handsomely bound in morocco by Hayday.

BANFIELD AND WELD.—THE STATISTICAL COMPANION,

Corrected to 1850; exhibiting the most interesting Facts in Moral and Intellectual, Vital, Economical, and Political Statistics, at home and abroad. Compiled from Official and other authentic Sources, by T. C. BANFIELD, Statistical Clerk to the Council of Education; and R. C. WELD Assistant-Secretary to the Royal Society. New Edition (1850), corrected and extended. Fcp. 8vo. 5s. cloth.

BAYLDON'S ART OF VALUING RENTS AND TILLAGES,

And Tenant's Right of Entering and Quitting Farms, explained by several Specimens of Valuations; with Remarks on the Cultivation pursued on Soils in different Situations. Adapted to the Use of Landlords, Land-Agents, Appraisers, Farmers, and Tenants. New Edition; corrected and revised by John Donaldson. 8vo. 10s. 6d. cloth.

BLACK.—A PRACTICAL TREATISE ON BREWING,

Based on Chemical and Economical Principles: with Formulæ for Public Brewers, and Instructions for Private Families. By WILLIAM BLACK, Practical Brewer. New Edition, with considerable Additions. 8vo. 10s. 6d. cloth.

BLAINE.—AN ENCYCLOPÆDIA OF RURAL SPORTS;

Or, a complete Account, Historical, Practical, and Descriptive, of Hunting, Shooting, Fishing Racing, and other Field Sports and Athletic Amusements of the present day. By DELABERE P. BLAINE, Esq. Author of "Canine Pathology," &c. Illustrated by nearly 600 Engravings on Wood, by R. Branston, from Drawings by Alken, T. Landseer, Dickes, &c. A New and thoroughly revised Edition, corrected to 1851. In One Large Volume, 8vo. [*In the press.*

BLAIR'S CHRONOLOGICAL AND HISTORICAL TABLES,

From the Creation to the present time: with Additions and Corrections from the most authentic Writers; including the Computation of St. Paul, as connecting the Period from the Exode to the Temple. Under the revision of Sir HENRY ELLIS, K.H., late Principal Librarian of the British Museum. Imperial 8vo. 31s. 6d. half-bound in morocco.

BLOOMFIELD.—THE GREEK TESTAMENT:

With copious English Notes, Critical, Philological, and Explanatory. Especially formed for the use of advanced Students and Candidates for Holy Orders. By Rev. S. T. BLOOMFIELD, D.D. F.S.A. New Edition. 2 vols. 8vo. with a Map of Palestine, £2, cloth.

THE REV. DR. S. T. BLOOMFIELD'S ADDITIONAL ANNO-

TATIONS, CRITICAL, PHILOLOGICAL, and EXPLANATORY, on the NEW TESTAMENT: being a Supplemental Volume to his Edition of *The Greek Testament with English Notes*, in 2 vols. 8vo. In One large Volume, of 460 pages, printed in double columns, uniformly with Dr. Bloomfield's larger Edition of the Greek Testament with English Notes; and so arranged as to be divisible into Two Parts, each of which may be bound up with the Volume to which it refers. 8vo. 15s. cloth.

BLOOMFIELD.—COLLEGE & SCHOOL GREEK TESTAMENT:

With shorter English Notes, Critical, Philological, and Explanatory, formed for use in Colleges and the Public Schools. By the Rev. S. T. BLOOMFIELD, D.D F.S.A. New Edition. greatly enlarged and improved. Fcp. 8vo. 10s. 6d. cloth.

BLOOMFIELD.—GREEK AND ENGLISH LEXICON TO THE

NEW TESTAMENT: especially adapted to the use of Colleges, and the Higher Classes in Public Schools; but also intended as a convenient Manual for Biblical Students in general By Dr. S. T. BLOOMFIELD. New Edition, enlarged and improved. Fcp. 8vo. 10s. 6d. cloth.

BOURNE.—A CATECHISM OF THE STEAM ENGINE,

Illustrative of the Scientific Principles upon which its Operation depends, and the Practical Details of its Structure, in its applications to Mines, Mills, Steam Navigation, and Railways: with various Suggestions of Improvement. By JOHN BOURNE, C.E., Editor of the Artisan Club's "Treatise on the Steam Engine." 3d Edition, corrected. Fcp. 8vo. 6s. cloth.

BRANDE.—A DICTIONARY OF SCIENCE, LITERATURE,

AND ART; comprising the History, Description, and Scientific Principles of every Branch of Human Knowledge; with the Derivation and Definition of all the Terms in General Use. Edited by W. T. BRANDE, F.R.S.L. & E.; assisted by Dr. J. CAUVIN. A New and thoroughly revised Edition, corrected to 1851. In One Large Volume, 8vo. with Wood Engravings.
[*In the press.*

BUDGE.—THE PRACTICAL MINER'S GUIDE.

Comprising a Set of Trigonometrical Tables adapted to all the purposes of Oblique or Diagonal, Vertical, Horizontal, and Traverse Dialling; with their application to the Dial, Exercise of Drifts, Lodes, Slides, Levelling, Inaccessible Distances, Heights, &c. By J. BUDGE. New Edition, considerably enlarged. 8vo. with Portrait of the Author, 12s. cloth.

BULL.—THE MATERNAL MANAGEMENT OF CHILDREN,
in HEALTH and DISEASE. By T. BULL, M.D. Member of the Royal College of Physicians; formerly Physician-Accoucheur to the Finsbury Midwifery Institution, and Lecturer on Midwifery and on the Diseases of Women and Children. New Edition, carefully revised and enlarged. Fcp. 8vo. 5s. cloth.

BULL.—HINTS TO MOTHERS,
For the Management of their Health during the Period of Pregnancy and in the Lying-in Room: with an Exposure of Popular Errors in connexion with those subjects, &c.; and Hints upon Nursing. By THOMAS BULL, M.D. New Edition, carefully revised and enlarged. Fcp. 8vo. 5s. cloth.

BUNSEN.—EGYPT'S PLACE IN UNIVERSAL HISTORY:
An Historical Investigation, in Five Books. By CHRISTIAN C. J. BUNSEN, D.Ph. & D.C.L. Translated from the German, by C. H. COTTRELL, Esq. M.A.—Vol. I. containing the First Book, or Sources and Primeval Facts of Egyptian History: with an Egyptian Grammar and Dictionary, and a complete List of Hieroglyphical Signs; an Appendix of Authorities, embracing the complete Text of Manetho and Eratosthenes, Ægyptiaca from Pliny, Strabo, &c.; and Plates representing the Egyptian Divinities. 8vo. with numerous illustrations, 28s. cloth.

BISHOP BUTLER'S SKETCH OF MODERN AND ANCIENT
GEOGRAPHY, for the use of Schools. An entirely New Edition (1851), carefully revised throughout, with such Alterations introduced as continually progressive Discoveries and the latest Information have rendered necessary. Edited by the Author's Son, the Rev. THOMAS BUTLER, Rector of Langar. 8vo. 9s. cloth.

BISHOP BUTLER'S GENERAL ATLAS OF MODERN AND
ANCIENT GEOGRAPHY; comprising Fifty-one full-coloured Maps; with complete Indices. New Edition (1851). nearly all re-engraved, enlarged, and greatly improved; with corrections from the most authentic sources in both the Ancient and Modern Maps, many of which are entirely new. Edited by the Author's Son, the Rev. T. BUTLER. Royal 4to. 24s. half-bound.

Separately { The Modern Atlas. 28 full-coloured Maps. Royal 8vo. 12s. half-bound.
{ The Ancient Atlas. 23 full-coloured Maps. Royal 8vo. 12s. half-bound.

THE CABINET LAWYER:
A Popular Digest of the Laws of England, Civil and Criminal; with a Dictionary of Law Terms, Maxims, Statutes, and Judicial Antiquities; Correct Tables of Assessed Taxes, Stamp Duties, Excise Licenses, and Post-Horse Duties; Post-Office Regulations, and Prison Discipline. 15th Edition (1851), enlarged, and corrected throughout, with the Legal Decisions and Statutes to Michaelmas Term, 13 and 14 Victoria. Fcp. 8vo. 10s. 6d. cloth.

CALLCOTT.—A SCRIPTURE HERBAL.
With upwards of 120 Wood Engravings. By Lady CALLCOTT. Square crown 8vo. 25s. cloth.

CATLOW.—POPULAR CONCHOLOGY;
Or, the Shell Cabinet arranged: being an Introduction to the Modern System of Conchology: with a sketch of the Natural History of the Animals, an account of the Formation of the Shells, and a complete Descriptive List of the Families and Genera. By AGNES CATLOW. Fcp. 8vo. with 312 Woodcuts, 10s. 6d. cloth.

CHESNEY.—THE EXPEDITION FOR THE SURVEY OF
THE RIVERS EUPHRATES and TIGRIS, carried on by order of the British Government, in the Years 1835, 1836, and 1837. By Lieut.-Col. CHESNEY, R.A., F.R.S., Commander of the Expedition. Vols. I. and II. in royal 8vo. with a coloured Index Map and numerous Plates and Woodcuts, 63s. cloth.—Also, an ATLAS of Thirteen Charts of the Expedition, price £1. 11s. 6d in case.

⁎ The entire work will consist of Four Volumes, royal 8vo. embellished with Ninety-seven Plates, besides numerous Woodcut Illustrations, from Drawings chiefly made by Officers employed in the Surveys.

JOHN COAD'S MEMORANDUM.—A CONTEMPORARY AC-
COUNT of the SUFFERINGS of the REBELS sentenced to TRANSPORTATION by JUDGE JEFFERIES ; being, A Memorandum of the Wonderful Providences of God to a poor unworthy Creature, during the time of the Duke of Monmouth's Rebellion, and to the Revolution in 1688. By JOHN COAD, one the Sufferers. Square fcp. 8vo. 4s. 6d. cloth.

CONYBEARE AND HOWSON.—THE LIFE AND EPISTLES
of SAINT PAUL; comprising a complete Biography of the Apostle, and a Paraphrastic Translation of his Epistles inserted in Chronological Order. Edited by the Rev. W. J. CONYBEARE, M.A. late Fellow of Trinity College, Cambridge; and the Rev. J. S. HOWSON, M.A. late Principal of the Collegiate Institution, Liverpool. Copiously illustrated by nu me-rous Engravings on Steel and Wood of the Principal Places visited by the Apostle, from Original Drawings made on the spot by W. H. Bartlett; and by Maps, Charts, Woodcuts of Coins, &c. Vol. I. Part I.; with Thirteen Engravings on Steel, Seven Maps and Plans, and numerous Woodcuts. 4to. 17s. boards.

. In course of publication in Twenty Parts, price 2s. each ; of which Twelve are now ready.

CONVERSATIONS ON BOTANY.
New Edition, improved. Fcp. 8vo. 22 Plates, 7s. 6d. cloth ; with the plates coloured, 12s. cloth.

CONVERSATIONS ON MINERALOGY.
With Plates, engraved by Mr. and Mrs. Lowry, from Original Drawings. New Edition, enlarged. 2 vols. fcp. 8vo. 14s. cloth.

COOK.—THE ACTS OF THE APOSTLES :
With a Commentary, and Practical and Devotional Suggestions, for the Use of Readers and Students of the English Bible. By the Rev. F. C. COOK, M.A. one of Her Majesty's Inspec-tors of Church Schools. Post 8vo. 8s. 6d. cloth.

COOPER.—PRACTICAL AND FAMILIAR SERMONS,
Designed for Parochial and Domestic Instruction. By the Rev. EDWARD COOPER. New Edi-tion. 7 vols. 12mo. £1. 18s. boards.

COPLAND.—A DICTIONARY OF PRACTICAL MEDICINE ;
comprising General Pathology, the Nature and Treatment of Diseases, Morbid Structures, and the Disorders especially incidental to Climates, to Sex, and to the different Epochs of Life; with numerous approved Formulæ of the Medicines recommended. By JAMES COPLAND, M.D. Consulting Physician to Queen Charlotte's Lying-in Hospital, &c. &c. Vols. I. and II. 8vo. £3, cloth ; and Parts X. to XIV. 4s. 6d. each, sewed.

THE CHILDREN'S OWN SUNDAY-BOOK.
By Miss JULIA CORNER, Author of " Questions on the History of Europe," &c. With Two Illustrations engraved on Steel. Square fcp. 8vo. 5s. cloth.

CRESY.—AN ENCYCLOPÆDIA OF CIVIL ENGINEERING,
Historical, Theoretical, and Practical. By EDWARD CRESY, F.S.A. C.E. In One very large Volume, illustrated by upwards of Three Thousand Engravings on Wood, explanatory of the Principles, Machinery, and Constructions which come under the Direction of the Civil Engineer. 8vo. £3. 13s. 6d. cloth.

THE CRICKET-FIELD ; OR, THE SCIENCE AND HISTORY
of the GAME. Illustrated with Diagrams, and enlivened with Anecdotes. By the Author of " Principles of Scientific Batting," " Recollections of College Days," &c. With Two Engravings on Steel; uniform with *Harry Hieover's Hunting-Field.* Fcp. 8vo.

[*Nearly ready.*

CROCKER'S ELEMENTS OF LAND SURVEYING.

New Edition, corrected throughout, and considerably improved and modernized, by T. G. Bunt, Land Surveyor. To which are added, TABLES OF SIX-FIGURE LOGARITHMS, &c., superintended by R. Farley, of the Nautical Almanac Establishment. Post 8vo. 12s. cloth.

*** Mr. Farley's Tables of Six-Figure Logarithms may be had separately, price 4s. 6d.

DALE.—THE DOMESTIC LITURGY AND FAMILY CHAP-

LAIN, in two Parts: The First Part being Church Services adapted for domestic use, with Prayers for every day of the week, selected exclusively from the Book of Common Prayer; Part II. comprising an appropriate Sermon for every Sunday in the year. By the Rev. Thomas Dale, M.A., Canon Residentiary of St. Paul's Cathedral. 2d Edition. Post 4to. 21s. cloth: or, bound by Hayday, 31s. 6d. calf lettered; £2. 10s. morocco.

Separately { The Family Chaplain, price 12s. cloth.
{ The Domestic Liturgy, price 10s. 6d. cloth.

DANDOLO.—THE ITALIAN VOLUNTEERS AND LOMBARD

RIFLE BRIGADE in the YEARS 1848-49. Translated from the Italian of Emilio Dandolo. Edited by the Rev. T. L. Wolley. [Nearly ready.

DELABECHE.—THE GEOLOGICAL OBSERVER.

By Sir Henry T. Delabeche, F.R.S., Director-General of the Geological Survey of the United Kingdom. In One large Volume, with many Wood Engravings. 8vo. 18s. cloth.

DELABECHE.—REPORT ON THE GEOLOGY OF CORN-

WALL, DEVON, and WEST SOMERSET. By Henry T. Delabeche, F.R.S., Director-General of the Geological Survey of the United Kingdom. Published by Order of the Lords Commissioners of H.M. Treasury. 8vo. with Maps, Woodcuts, and 12 large Plates, 14s. cloth.

DE LA RIVE'S WORK ON ELECTRICITY.—A TREATISE

on ELECTRICITY; its Theory and Practical Application. By A. De la Rive, of the Academy of Geneva. Illustrated with numerous Wood Engravings. 2 vols. 8vo.
[Ne rly ready.

DENNISTOUN.—MEMOIRS OF THE DUKES OF URBINO;

Illustrating the Arms, Arts, and Literature of Italy, from MCCCCXL. to MDCXXX. By James Dennistoun, of Dennistoun. With numerous Portraits, Plates, Fac-similes, and Engravings on Wood. 3 vols. crown 8vo. £2. 8s. cloth.

"The object of these volumes is to combine a general picture of the progress of Italian literature and art under the patronage of the smaller Italian principalities, with a history of the houses of Montefeltro and Della Rovere, so far as they were connected with Urbino. For the execution of his purpose Mr. Dennistoun enjoyed numerous advantages. He has resided in Italy for many years, and is well acquainted with the country whose history as a dukedom he intended to relate: besides the common research in printed volumes, he has gained access to various Italian libraries, including that of Urbino, and examined their manuscripts; and he has closely inspected Umbrian art, whether in palatial, military, or ecclesiastical architecture, or in painting. With the appreciation of Italy which such a course of study stimulates and implies, Mr. Dennistoun has good taste, a sound, though not always an unbiassed judgment, and a zeal for mediaeval subjects, especially art, almost enthusiastic." Spectator.

DISCIPLINE.

By the Authoress of "Letters to my Unknown Friends," "Twelve Years Ago," "Some Passages from Modern History," and "Letters on Happiness." Second Edition, enlarged. 18mo. 2s. 6d. cloth.

DIXON.—THE GEOLOGY AND FOSSILS OF THE TERTIARY

and CRETACEOUS FORMATIONS of SUSSEX. By the late Frederick Dixon, Esq. F.G.S. The Fossils engraved from Drawings by Messrs. Sowerby, Dinkel, and Erxleben. In One large Volume, with 44 Plates and many Wood Engravings. Royal 4to. 63s. cloth; India Proofs, £5s. 5s.

*** In this work are embodied the results of many years' Geological and Palæontological observations by the Author, together with some remarks on the Archæology of Sussex. It also includes Descriptions—Of the fossil Reptilia, by Prof. Owen, F.R.S.; of the Echinoderma, by Prof. Edward Forbes, F.R.S.; of the Crustacea, by Prof. Thomas Bell, Sec. R.S.; of the Corals, by William Lonsdale, Esq. F.G.S.; and of the fossil Shells, by J. De Carle Sowerby, Esq. F.L.S.

DOUBLEDAY AND HEWITSON'S BUTTERFLIES. — THE
GENERA of DIURNAL LEPIDOPTERA; comprising their Generic Characters—a Notice of the Habits and Transformations—and a Catalogue of the Species of each Genus. By EDWARD DOUBLEDAY, Esq. F.L.S. &c., late Assistant in the Zoological Department of the British Museum. Continued by J. O. WESTWOOD, Esq. Illustrated with 75 Coloured Plates, by W. C. HEWITSON, Esq. Author of " British Oology." Imperial 4to. uniform with Gray and Mitchell's " Genera of Birds."

₊ In course of publication in Monthly Parts, 5s. each ; of which 41 have appeared. The publication, which had been suspended in consequence of the death of Mr. Doubleday, is now resumed, and will be continued regularly until the completion of the work in about Fifty Parts.

DRESDEN GALLERY.—THE MOST CELEBRATED PICTURES
of the ROYAL GALLERY at DRESDEN, drawn on Stone, from the Originals, by Franz Hanfstaengl : with Descriptive and Biographical Notices, in French and German. Nos. I. to LVIII. imperial folio, each containing 3 Plates, with Letter-press, price 20s. to Subscribers ; to Non Subscribers, 30s. Single Plates, 12s. each.

₊ To be completed in 2 more numbers, price 20s. each, to Subscribers. Nos. LI. to LX. contain each *Four Plates* and Letterpress.

DUNLOP.—THE HISTORY OF FICTION :
Being a Critical Account of the most celebrated Prose Works of Fiction, from the earliest Greek Romances to the Novels of the Present Age. By JOHN DUNLOP, Esq. New Edition, complete in One Volume. 8vo. 15s. cloth.

EASTLAKE. — MATERIALS FOR A HISTORY OF OIL
PAINTING. By CHARLES LOCK EASTLAKE, Esq. P.R.A. F.R.S. F.S.A.; Secretary to the Royal Commission for Promoting the Fine Arts in connexion with the rebuilding of the Houses of Parliament, &c. 8vo. 16s. cloth.

₊ Vol. II. On the Italian Practice of Oil Painting, is *preparing for publication.*

ELMES'S THOUGHT BOOK, OR HORÆ VACIVÆ.
Horæ Vacivæ ; or, a Thought Book of the Wise Spirits of all Ages and all Countries, for all Men and all Hours. Collected, arranged, and edited by JAMES ELMES, Author of " Memoirs of Sir Christopher Wren," &c. Fcp. 16mo. (printed by C. Whittingham, Chiswick), 4s. 6d. bound in cloth.

THE ENGLISHMAN'S GREEK CONCORDANCE OF THE
NEW TESTAMENT: being an Attempt at a Verbal Connexion between the Greek and the English Texts ; including a Concordance to the Proper Names, with Indexes, Greek-English and English-Greek. New Edition, with a new Index. Royal 8vo. 42s. cloth.

THE ENGLISHMAN'S HEBREW AND CHALDEE CON-
RDANCE of the OLD TESTAMENT; being an Attempt at a Verbal Connexion between the Original and the English Translations : with Indexes, a List of the Proper Names and their occurrences, &c. 2 vols. royal 8vo. £3. 13s. 6d. cloth ; large paper, £4. 14s. 6d.

EPHEMERA.—THE BOOK OF THE SALMON :
In Two Parts. Part I. The Theory, Principles, and Practice of Fly-Fishing for Salmon : with Lists of good Salmon Flies for every good River in the Empire ; Part II. The Natural History of the Salmon, all its known Habits described, and the best way of artificially Breeding It explained. Usefully illustrated with numerous Coloured Engravings of Salmon Flies and Salmon Fry. By EPHEMERA, Author of " A Hand-Book of Angling ;" assisted by ANDREW YOUNG, of Invershin, Manager of the Duke of Sutherland's Salmon Fisheries. Fcp. 8vo. with coloured Plates, 14s. cloth.

EPHEMERA.—A HAND-BOOK OF ANGLING ;
Teaching Fly-fishing, Trolling, Bottom-fishing, Salmon-fishing ; with the Natural History of River Fish, and the best modes of Catching them. By EPHEMERA, of " Bell's Life in London." New Edition, enlarged. Fcp 8vo. with numerous Woodcuts, 9s. cloth.

ERMAN.—TRAVELS IN SIBERIA:

including Excursions northwards, down the Obi, to the Polar Circle, and southwards to the Chinese Frontier. By ADOLPH ERMAN. Translated by W. D. COOLEY, Esq. Author of "The History of Maritime and Inland Discovery;" 2 vols. 8vo. with Map, 31s. 6d. cloth.

EVANS.—THE SUGAR PLANTER'S MANUAL:

Being a Treatise on the Art of obtaining Sugar from the Sugar Cane. By W. J. EVANS, M.D. 8vo. 9s. cloth.

FORBES.—DAHOMEY AND THE DAHOMANS:

Being the Journals of Two Missions to the King of Dahomey, and Residence at his Capital, in the Years 1849 and 1850. By FREDERICK E. FORBES, Commander, R.N., F.R.G.S.; Author of "Five Years in China," and "Six Months in the African Blockade." With 10 Plates, printed in Colours, and 8 Wood Engravings. 2 vols. post 8vo. 21s. cloth.

FORESTER AND BIDDULPH'S NORWAY.

Norway in 1848 and 1849: containing Rambles among the Fields and Fjords of the Central and Western Districts; and including Remarks on its Political, Military, Ecclesiastical, and Social Organisation. By THOMAS FORESTER, Esq. With Extracts from the Journals of Lieutenant M. S. BIDDULPH, Royal Artillery. With a new Map, Woodcuts, and Ten coloured Plates from Drawings made on the spot. 8vo. 18s. cloth.

FOSS.—THE JUDGES OF ENGLAND:

with Sketches of their Lives, and Miscellaneous Notices connected with the Courts at Westminster from the time of the Conquest. By EDWARD FOSS, F.S.A. of the Inner Temple. Vols. I. and II. 8vo. 28s. cloth.

FOSTER.—A HANDBOOK OF MODERN EUROPEAN LITE-

RATURE: British, Danish, Dutch, French, German, Hungarian, Italian, Polish and Russian, Portuguese, Spanish, and Swedish. With a full Biographical and Chronological Index. By Mrs. FOSTER. Fcp. 8vo. 8s. 6d. cloth.

*** The object of this book is, not so much to give elaborate criticisms on the various writers in the language to whose literature it is intended as a guide, as to direct the student to the best writers in each, and to inform him on what subjects they have written.

GIBBON'S HISTORY OF THE DECLINE AND FALL OF THE

ROMAN EMPIRE. A new Edition, complete in One Volume. With an Account of the Author's Life and Writings, by ALEXANDER CHALMERS, Esq. F.A.S. 8vo. with Portrait, 18s. cloth.

*** An Edition, in 8 vols. 8vo. 60s. boards.

GILBART.—A PRACTICAL TREATISE ON BANKING.

By JAMES WILLIAM GILBART, F.R.S. General Manager of the London and Westminster Bank. 5th Edition, with Portrait of the Author, and View of the "London and Westminster Bank," Lothbury. 2 vols. 8vo. 24s. cloth.

GOLDSMITH. — THE POETICAL WORKS OF OLIVER

GOLDSMITH. Illustrated by Wood Engravings, from Designs by Members of the Etching Club. With a Biographical Memoir, and Notes on the Poems. Edited by BOLTON CORNEY, Esq. Square crown 8vo. uniform with *Thomson's Seasons illustrated by the Etching Club*, 21s. cloth; or, bound in morocco by Hayday, £1. 16s.

GOSSE.—NATURAL HISTORY OF THE ISLAND OF JAMAICA.

By P. H. GOSSE, Author of "The Birds of Jamaica," "Popular British Ornithology," &c. Post 8vo. with coloured Plates.   [*Nearly ready.*

GOWER.—THE SCIENTIFIC PHÆNOMENA OF DOMESTIC

LIFE, familiarly explained. By CHARLES FOOTE GOWER. New Edition. Fcp. 8vo. with Wood Engravings, 5s. cloth.

c

GRAHAM.—ENGLISH; OR, THE ART OF COMPOSITION:
explained in a Series of Instructions and Examples. By G. F. GRAHAM. New Edition, revised and improved. Fcp. 8vo. 6s. cloth.

GRANT.—LETTERS FROM THE MOUNTAINS.
Being the Correspondence with her Friends, between the years 1773 and 1803. By Mrs GRANT, of Laggan. New Edition. Edited, with Notes and Additions, by her son, J. P. GRANT, Esq. 2 vols. post 8vo. 21s. cloth.

GRANT.— MEMOIR AND CORRESPONDENCE OF THE
late Mrs. Grant, of Laggan, Author of "Letters from the Mountains," "Memoirs of an American Lady," &c. Edited by her Son, J. P. GRANT, Esq. New Edition. 3 vols. post 8vo. with Portrait, 21s. 6d. cloth.

GRAY.—TABLES AND FORMULÆ FOR THE COMPUTATION
of LIFE CONTINGENCIES; with copious Examples of Annuity, Assurance, and Friendly Society Calculations. By PETER GRAY, F.R.A.S. Associate of the Institute of Actuaries of Great Britain and Ireland. Royal 8vo. 15s. cloth.

GRAY AND MITCHELL'S ORNITHOLOGY.—THE GENERA
Of BIRDS; comprising their Generic Characters, a Notice of the Habits of each Genus, and an extensive List of Species, referred to their several Genera. By GEORGE ROBERT GRAY, Acad. Imp. Georg. Florent. Soc. Corresp., Senior Assistant of the Natural History Department in the British Museum. Illustrated with 360 Plates (175 plain and 185 coloured), drawn on stone, by DAVID WILLIAM MITCHELL, B.A., F.L.S., Secretary to the Zoological Society of London, &c. 3 vols. imperial 4to. £31. 10s. half-bound morocco, gilt tops.

GWILT.—AN ENCYCLOPÆDIA OF ARCHITECTURE;
Historical, Theoretical, and Practical. By JOSEPH GWILT. Illustrated with more than One Thousand Engravings on Wood, from Designs by J. S. GWILT. Second Edition (1851), with a Supplemental View of the Symmetry and Stability of Gothic Architecture; comprising upwards of Eighty additional Woodcuts. 8vo. 52s. 6d. cloth.

SUPPLEMENT TO GWILT'S ENCYCLOPÆDIA OF ARCHI-
TECTURE. Comprising a View of the Symmetry and Stability of Gothic Architecture; Addenda to the Glossary; and an Index to the entire Work. By JOSEPH GWILT. Illustrated by upwards of Eighty Wood Engravings by R. Branston. 8vo. 6s. cloth.

SIDNEY HALL'S NEW GENERAL LARGE LIBRARY ATLAS
OF FIFTY-THREE MAPS (size 20 in. by 16 in.), with the Divisions and Boundaries carefully coloured; and an Alphabetical Index of all the Names contained in the Maps, with their Latitude and Longitude. An entirely New Edition, corrected throughout from the best and most recent Authorities; with all the Railways laid down, and many of the Maps re-drawn and re-engraved. Colombier 4to. £5. 5s. half-bound in russia.

SIDNEY HALL'S RAILWAY MAP OF ENGLAND AND
WALES. Square fcp. 8vo. 2s. 6d. cloth.

. The Map of England and Wales, contained in "Sidney Hall's Large Railway Atlas" (size 20 in. by 16 in.) corrected and re-engraved, with all the Lines of Railway laid down, may be had separately, price 2s. 6d., coloured and mounted on folding canvas in a case for the pocket.

HAMILTON.—CRITICAL ESSAYS ON PHILOSOPHY, LITE-
RATURE, and ACADEMICAL REFORM, contributed to The Edinburgh Review by Sir William Hamilton, Bart. With additional Notes and Appendices. [In the press.

HARRISON.—ON THE RISE, PROGRESS, AND PRESENT
STRUCTURE of the ENGLISH LANGUAGE. By the Rev. M. HARRISON, M.A. late Fellow of Queen's College, Oxford. Post 8vo. 8s. 6d. cloth.

HARRY HIEOVER.—THE HUNTING-FIELD.

By HARRY HIEOVER, Author of "Stable-Talk and Table-Talk; or, Spectacles for Young Sportsmen." With Two Plates—One representing *The Right Sort;* the other, *The Wrong Sort.* Fcp. 8vo. 5s. half-bound.

HARRY HIEOVER.—PRACTICAL HORSEMANSHIP.

By HARRY HIEOVER, Author of "Stable Talk and Table Talk; or, Spectacles for Young Sportsmen." With 2 Plates—One representing *Going like Workmen;* the other, *Going like Muffs.* Fcp. 8vo. 5s. half-bound.

HARRY HIEOVER.—THE STUD, FOR PRACTICAL PUR-

POSES AND PRACTICAL MEN: being a Guide to the Choice of a Horse for use more than for show. By HARRY HIEOVER, Author of "Stable Talk and Table Talk." With Two Plates —One representing *A pretty good sort for most purposes;* the other, *'Rayther' a bad sort for any purpose.* Fcp. 8vo. 5s. half-bound.

HARRY HIEOVER.—THE POCKET AND THE STUD;

Or, Practical Hints on the Management of the Stable. By HARRY HIEOVER, Author of "Stable-Talk and Table-Talk; or, Spectacles for Young Sportsmen." Second Edition; with Portrait of the Author on his favourite Horse *Harlequin.* Fcp. 8vo. 5s. half-bound.

HARRY HIEOVER.—STABLE TALK AND TABLE TALK;

or, SPECTACLES for YOUNG SPORTSMEN. By HARRY HIEOVER. New Edition. 2 vols. 8vo. with Portrait, 24s. cloth.

HAWKER.—INSTRUCTIONS TO YOUNG SPORTSMEN

In all that relates to Guns and Shooting. By Lieut.-Col. P. HAWKER. New Edition, corrected, enlarged, and improved; with Eighty-five Plates and Woodcuts by Adlard and Branston, from Drawings by C. Varley, Dickes, &c. 8vo. 21s. cloth.

HAYDN.—THE BOOK OF DIGNITIES; OR, ROLLS OF THE

OFFICIAL PERSONAGES of the BRITISH EMPIRE, from the EARLIEST PERIODS to the PRESENT TIME: comprising the Administrations of Great Britain; the Offices of State, and all the Public Departments; the Ecclesiastical Dignitaries; the Functionaries of the Law; the Commanders of the Army and Navy; and the Hereditary Honours and other Distinctions conferred upon Families and Public Men. Being a New Edition, improved and continued, of BEATSON'S POLITICAL INDEX. By JOSEPH HAYDN, Compiler of "The Dictionary of Dates," and other Works. In One very large Volume, 8vo. [*In the Spring.*

HEAD.—THE METAMORPHOSES OF APULEIUS:

A Romance of the Second Century. Translated from the Latin by Sir GEORGE HEAD, Author of "A Tour of Many Days in Rome;" Translator of "Historical Memoirs of Cardinal Pacca." Post 8vo. 12s. cloth.

HEAD.—HISTORICAL MEMOIRS OF CARDINAL PACCA,

Prime Minister to Pius VII. Written by Himself. Translated from the Italian, by Sir GEORGE HEAD, Author of "Rome: a Tour of Many Days." 2 vols. post 8vo. 21s. cloth.

SIR GEORGE HEAD.—ROME:

A Tour of Many Days. By Sir GEORGE HEAD. 3 vols. 8vo. 36s. cloth.

SIR JOHN HERSCHEL.—OUTLINES OF ASTRONOMY.

By Sir JOHN F. W. HERSCHEL, Bart. &c. &c. &c. New Edition; with Plates and Wood Engravings. 8vo. 18s. cloth.

MRS. HEY.—THE MORAL OF FLOWERS;

Or, Thoughts gathered from the Field and the Garden. By Mrs. Hey. Being a New Edition of "The Moral of Flowers ;" and consisting of Poetical Thoughts on Garden and Field Flowers, accompanied by Drawings beautifully coloured after Nature. Square crown 8vo. uniform in size with *Thomson's Seasons illustrated by the Etching Club*, 31s. cloth.

MRS. HEY.—SYLVAN MUSINGS;

Or, the Spirit of the Woods. By Mrs. Hey. Being a New Edition of the "Spirit of the Woods ;" and consisting of Poetical Thoughts on Forest Trees, accompanied by Drawings of Blossoms and Foliage, beautifully coloured after Nature. Square crown 8vo. uniform in size with *Thomson's Seasons illustrated by the Etching Club*, 31s. cloth.

HINTS ON ETIQUETTE AND THE USAGES OF SOCIETY:

With a Glance at Bad Habits. By Ἀγωγός. "Manners make the man." New Edition, revised (with additions) by a Lady of Rank. Fcp. 8vo. 2s. 6d. cloth.

HOARE.—A PRACTICAL TREATISE ON THE CULTIVATION

OF THE GRAPE VINE ON OPEN WALLS. By Clement Hoare. New Edition. 8vo. 7s. 6d. cloth.

LORD HOLLAND'S FOREIGN REMINISCENCES.—FOREIGN

REMINISCENCES. By Henry Richard Lord Holland. Comprising Anecdotes, and an Account of such Persons and Political Intrigues in Foreign Countries as have fallen within his Lordship's observation. Edited by his Son, Henry Edward Lord Holland; with Fac-simile. Post 8vo. 10s. 6d.

HOOK.—THE LAST DAYS OF OUR LORD'S MINISTRY:

A Course of Lectures on the principal Events of Passion Week. By Walter Farquhar Hook, D.D. Vicar of Leeds, Prebendary of Lincoln, and Chaplain in Ordinary to the Queen. New Edition. Fcp. 8vo. 6s. cloth.

HOOKER.—KEW GARDENS;

Or, a Popular Guide to the Royal Botanic Gardens of Kew. By Sir William Jackson Hooker, K.H. D.C.L. F.R.A. & L.S. &c. &c. Director. New Edition. 16mo. with numerous Wood Engravings, 6d. sewed.

HOOKER AND ARNOTT.—THE BRITISH FLORA;

Comprising the Phænogamous or Flowering Plants, and the Ferns. The Sixth Edition (1850), with Additions and Corrections; and numerous Figures illustrative of the Umbelliferous Plants, the Composite Plants, the Grasses, and the Ferns. By Sir W. J. Hooker, F.R.A. and L.S. &c., and G. A. Walker-Arnott, LL.D. F.L.S. and R.S. Ed ; Regius Professor of Botany in the University of Glasgow. In One very thick Volume, 12mo. with 12 Plates, 14s. cloth; or with the Plates coloured, price 21s.

HORNE.—AN INTRODUCTION TO THE CRITICAL STUDY

and KNOWLEDGE of the HOLY SCRIPTURES. By Thomas Hartwell Horne, B.D. of St. John's College, Cambridge; Rector of the united Parishes of St. Edmund the King and Martyr, and St. Nicholas Acons, Lombard Street ; Prebendary of St. Paul's. New Edition, revised and corrected. 5 vols. 8vo. with numerous Maps and Facsimiles of Biblical Manuscripts, 63s. cloth; or £5, bound in calf.

HORNE.—A COMPENDIOUS INTRODUCTION TO THE

STUDY of the BIBLE. By Thomas Hartwell Horne, B.D. of St. John's College, Cambridge. Being an Analysis of his "Introduction to the Critical Study and Knowledge of the Holy Scriptures." New Edition, corrected and enlarged. 12mo. with Maps and other Engravings, 9s. boards.

HOWITT.—THE CHILDREN'S YEAR.

By Mary Howitt. With Four Illustrations, engraved by John Absolon, from Original Designs by Anna Mary Howitt. Square 16mo. 5s. cloth.

HOWITT.—THE BOY'S COUNTRY BOOK:

Being the real Life of a Country Boy, written by himself; exhibiting all the Amusements, Pleasures, and Pursuits of Children in the Country. Edited by WILLIAM HOWITT. New Edition. Fcp. 8vo. with 40 Woodcuts, 6s. cloth.

HOWITT.—THE RURAL LIFE OF ENGLAND.

By WILLIAM HOWITT. New Edition, corrected and revised; with Engravings on wood, by Bewick and Williams: uniform with *Visits to Remarkable Places.* Medium 8vo. 21s. cloth.

HOWITT.—VISITS TO REMARKABLE PLACES;

Old Halls, Battle-Fields, and Scenes illustrative of Striking Passages in English History and Poetry. By WILLIAM HOWITT. New Edition; with 40 Engravings on Wood. Medium 8vo. 21s. cloth.

SECOND SERIES, chiefly in the Counties of NORTHUMBERLAND and DURHAM, with a Stroll along the BORDER. With upwards of 40 Engravings on Wood. Medium 8vo. 21s. cloth.

HOWSON.—SUNDAY EVENING:

Twelve Short Sermons for Family Reading. 1. The Presence of Christ; 2. Inward and Outward Life; 3. The Threefold Warning; 4. Our Father's Business; 5. Spiritual Murder; 6. The Duty of Amiability; 7. Honesty and Candour; 8. St. Peter and Cornelius; 9. The Midnight Traveller; 10. St. Andrew; 11. The Grave of Lazarus; 12. The Resurrection of the Body. By the Rev. J. S. HOWSON, M.A. Principal of the Collegiate Institution, Liverpool, and Chaplain to the Duke of Sutherland. Fcp. 8vo. 2s. 6d. cloth.

HOWSON AND CONYBEARE.—THE LIFE AND EPISTLES

of SAINT PAUL. By the Rev. J. S. HOWSON, M.A., and the Rev. W. J. CONYBEARE, M.A. 2 vols. 4to. very copiously illustrated by W. H. Bartlett.   [*See page* 6.

HUDSON.—THE EXECUTOR'S GUIDE.

By J. C. HUDSON, Esq. late of the Legacy Duty Office, London; Author of "Plain Directions for Making Wills," and "The Parent's Hand-book." New Edition. Fcp. 8vo. 5s. cloth.

HUDSON.—PLAIN DIRECTIONS FOR MAKING WILLS

In Conformity with the Law, and particularly with reference to the Act 7 Will. 4 and 1 Vict. c. 26. To which is added, a clear Exposition of the Law relating to the distribution of Personal Estate in the case of Intestacy; with two Forms of Wills, and much useful information, &c. By J. C. HUDSON, Esq. New Edition, corrected. Fcp. 8vo. 2s. 6d. cloth.

*** These Two works may be had in One Volume, 7s. cloth.

HUMBOLDT.—ASPECTS OF NATURE

In Different Lands and Different Climates; with Scientific Elucidations. By ALEXANDER VON HUMBOLDT. Translated, with the Author's sanction and co-operation, and at his express desire, by Mrs. SABINE. New Edition. 16mo. 6s. cloth: or in 2 vols. 3s. 6d. each, cloth; Half-a-Crown each, sewed.

BARON HUMBOLDT'S COSMOS;

Or, a Sketch of a Physical Description of the Universe. Translated, with the Author's sanction and co-operation, under the superintendence of Lieutenant-Colonel EDWARD SABINE, R.A. For. Sec. R.S. New Edition. Vols. I. and II. 16mo. Half-a-Crown each, sewed; 3s. 6d. each, cloth: or in post 8vo, 12s. each, cloth.—Vol. III. Part I. post 8vo. 6s. cloth: or in 16mo. 2s. 6d. sewed; 3s. 6d. cloth.

HUMPHREYS.—SENTIMENTS & SIMILES OF SHAKSPEARE:

A Classified Selection of Similes, Definitions, Descriptions, and other remarkable Passages in Shakspeare's Plays and Poems. With an elaborately illuminated border in the characteristic style of the Elizabethan Period, and other Embellishments. Bound in very massive carved and pierced covers containing in deep relief a medallion Head of Shakspeare. The Illuminations and Ornaments designed and executed by Henry Noel Humphreys, Illuminator of "A Record of the Black Prince," &c. Square post 8vo.   [*Nearly ready.*

HUMPHREYS.—A RECORD OF THE BLACK PRINCE;

Being a Selection of such Passages in his Life as have been most quaintly and strikingly narrated by the Chroniclers of the Period. Embellished with highly-wrought Miniatures and Borderings, selected from various Illuminated MSS. referring to Events connected with English History. By HENRY NOEL HUMPHREYS. Post 8vo. in a richly carved and pierced binding, 21s.

HUMPHREYS.—THE BOOK OF RUTH.

From the Holy Scriptures. Embellished with brilliant coloured Borders, selected from some of the finest Illuminated MSS. in the British Museum, the Bibliothèque Nationale, Paris, the Soane Museum, &c.; and with highly-finished Miniatures. The Illuminations executed by HENRY NOEL HUMPHREYS. Square fcp. 8vo. in deeply embossed leather covers, 21s.

HUMPHREYS. — MAXIMS AND PRECEPTS OF THE

SAVIOUR: being a Selection of the most beautiful Christian Precepts contained in the Four Gospels. Illustrated by a series of Illuminations of original character, founded on the Passages—"Behold the Fowls of the Air," &c., "Consider the Lilies of the Field," &c. The Illuminations executed by HENRY NOEL HUMPHREYS. Square fcp. 8vo. 21s. richly bound in stamped calf; or 30s. in morocco by Hayday.

HUMPHREYS.—THE MIRACLES OF OUR SAVIOUR.

With rich and appropriate Borders of original Design, a series of Illuminated Figures of the Apostles from the Old Masters, Six Illuminated Miniatures, and other Embellishments. The Illuminations executed by HENRY NOEL HUMPHREYS. Square fcp. 8vo. in massive carved covers, 21s.; or bound in morocco by Hayday. 30s.

HUMPHREYS.—PARABLES OF OUR LORD.

Richly illuminated with appropriate Borders printed in Colours and in Black and Gold; with a Design from one of the early German Engravers. The illuminations executed by HENRY NOEL HUMPHREYS. Square fcp. 8vo. 21s. in a massive carved binding; or 30s. bound in morocco by Hayday.

HUMPHREYS AND JONES.—THE ILLUMINATED BOOKS

OF THE MIDDLE AGES: A series of Fac-similes from the most beautiful MSS. of the Middle Ages, printed in Gold, Silver, and Colours by OWEN JONES; selected and described by HENRY NOEL HUMPHREYS. Elegantly bound in antique calf. Royal folio, £10. 10s.; imperial folio (large paper), £16. 16s.

HUNT.—RESEARCHES ON LIGHT :

An Examination of all the Phenomena connected with the Chemical and Molecular Changes produced by the Influence of the Solar Rays: embracing all the known Photographic Processes, and new Discoveries in the Art By ROBERT HUNT, Keeper of Mining Records, Museum of Practical Geology. 8vo. with Plate and Woodcuts, 10s. 6d. cloth.

MRS. JAMESON'S LEGENDS OF THE MONASTIC ORDERS,

as represented in the Fine Arts. Containing St. Benedict and the Early Benedictines in Italy, France, Spain, and Flanders; the Benedictines in England and in Germany; the Reformed Benedictines; early Royal Saints connected with the Benedictine Order; the Augustines; Orders derived from the Augustine Rule; the Mendicant Orders; the Jesuits; and the Order of the Visitation of St. Mary. Forming the SECOND SERIES of Sacred and Legendary Art. With Eleven Etchings by the Author, and 84 Woodcuts. Square crown 8vo. 28s. cloth.

MRS. JAMESON'S SACRED AND LEGENDARY ART ;

Or, Legends of the Saints and Martyrs. FIRST SERIES. Containing Legends of the Angels and Archangels; the Evangelists and Apostles; the Greek and Latin Fathers; the Magdalene; the Patron Saints; the Virgin Patronesses; the Martyrs; the Bishops; the Hermits; and the Warrior-Saints of Christendom. Second Edition (1850), printed in One Volume for the convenience of Students and Travellers; with numerous Woodcuts, and Sixteen Etchings by the Author. Square crown 8vo. 28s. cloth.

MRS. JAMESON'S LEGENDS OF THE MADONNA,

As represented in the Fine Arts. Forming the THIRD and *concluding* SERIES of *Sacred and Legendary Art.* By Mrs. JAMESON, Author of "Characteristics of Women," &c. With Etchings by the Author, and Engravings on Wood. Square crown 8vo.   [*In the press.*

JARDINE.—A TREATISE OF EQUIVOCATION ;

Wherein is largely discussed the question Whether a Catholicke or any other Person before a Magistrate, being demanded upon his Oath whether a Preiste were in such a place, may (notwithstanding his perfect knowledge to the contrary), without Perjury, and securely in conscience, answer No : with this secret meaning reserved in his Mynde—That he was not there, so that any man is bounde to detect it. Edited from the Original Manuscript in the Bodleian Library, by DAVID JARDINE, of the Middle Temple, Esq., Barrister at Law ; Author of the "Narrative of the Gunpowder Treason," prefixed to his edition of the "Criminal Trials."
[*In the press.*

JEFFREY. — CONTRIBUTIONS TO THE EDINBURGH

REVIEW. By FRANCIS JEFFREY, late One of the Judges of the Court of Session in Scotland. Second Edition. 3 vols. 8vo. 42s. cloth.

BISHOP JEREMY TAYLOR'S ENTIRE WORKS :

With the Life by Bishop HEBER. Revised and corrected by the Rev. CHARLES PAGE EDEN, Fellow of Oriel College, Oxford. Vols. II. III. IV. V. VI. VII. and VIII. 8vo. 10s. 6d. each.

* In course of publication, in Ten Volumes, price Half-a-Guinea each.—Vol. I. (the *last* in order of *publication*) will contain Bishop Heber's Life of Jeremy Taylor, extended by the Editor.—Vol. IX. is *in the press.*

BISHOP JEREMY TAYLOR.—READINGS FOR EVERY DAY

in LENT : compiled from the Writings of BISHOP JEREMY TAYLOR. By the Author of " Amy Herbert," "The Child's First History of Rome," &c. Fcp. 8vo. 5s. cloth.

JOHNSON.—THE FARMER'S ENCYCLOPÆDIA,

And Dictionary of Rural Affairs: embracing all the recent Discoveries in Agricultural Chemistry ; adapted to the comprehension of unscientific readers. By CUTHBERT W. JOHNSON, Esq. F.R.S. Barrister-at-Law ; Editor of the "Farmer's Almanack," &c. 8vo. with Wood Engravings, £2. 10s. cloth.

JOHNSON.—THE WISDOM OF THE RAMBLER, ADVEN-

TURER, and IDLER : comprising a Selection of 110 of the best Essays. By SAMUEL JOHNSON, LL.D. Fcp. 8vo. 7s. cloth.

JOHNSTON.—A NEW DICTIONARY OF GEOGRAPHY,

Descriptive, Physical, Statistical, and Historical: forming a complete General Gazetteer of the World. By ALEXANDER KEITH JOHNSTON, F.R.S.E. F.R.G.S. F.G.S. ; Geographer at Edinburgh in Ordinary to Her Majesty ; Author of "The Physical Atlas of Natural Phænomena." In One very large Volume of 1,440 pages ; comprising nearly Fifty Thousand Names of Places. 8vo. 36s. cloth ; or strongly half-bound in russia, with flexible back, price 41s.

KAY.—THE SOCIAL CONDITION AND EDUCATION OF

the PEOPLE in ENGLAND and EUROPE : shewing the Results of the Primary Schools and of the Division of Landed Property in Foreign Countries. By JOSEPH KAY, Esq. M.A. of Trinity College, Cambridge: Barrister-at-Law ; and late Travelling Bachelor of the University of Cambridge. 2 thick vols. post 8vo. 21s. cloth.

KEMBLE.—THE SAXONS IN ENGLAND :

a History of the English Commonwealth till the period of the Norman Conquest. By JOHN MITCHELL KEMBLE, M.A., F.C.P.S., &c. 2 vols. 8vo. 28s. cloth.

KINDERSLEY.—THE VERY JOYOUS, PLEASANT, AND

REFRESHING HISTORY of the Feats, Exploits, Triumphs, and Achievements of the Good Knight, without Fear and without Reproach, the gentle LORD DE BAYARD. Set forth in English by EDWARD COCKBURN KINDERSLEY, Esq. With Ornamental Headings, and Frontispiece by R. H. Wehnert. Square post 8vo. 9s. 6d. cloth.

KIRBY & SPENCE.—AN INTRODUCTION TO ENTOMOLOGY;

Or, Elements of the Natural History of Insects : comprising an account of noxious and useful Insects, of their Metamorphoses, Food, Stratagems, Habitations, Societies, Motions, Noises, Hybernation, Instinct, &c. By W. KIRBY, M.A. F.R.S. & L.S. Rector of Barham; and W. SPENCE, Esq. F.R.S. & L.S. New Edition. 2 vols. 8vo. with Plates, 31s. 6d. cloth.

LAING.—OBSERVATIONS ON THE SOCIAL AND POLI-

TICAL STATE of the EUROPEAN PEOPLE in 1848 and 1849 : being the Second Series of " Notes of a Traveller." By SAMUEL LAING, Esq. Author of " A Journal of a Residence in Norway," " A Tour in Sweden," the Translation of " The Heimskringla," and of " Notes of a Traveller on the Social and Political State of France, Prussia, &c." 8vo. 14s. cloth.

LATHAM.—ON DISEASES OF THE HEART.

Lectures on Subjects connected with Clinical Medicine; comprising Diseases of the Heart. By P. M. LATHAM, M. D., Physician Extraordinary to the Queen; and late Physician to St. Bartholomew's Hospital. New Edition. 2 vols. 12mo. 16s. cloth.

LEE.—ELEMENTS OF NATURAL HISTORY;

Or, First Principles of Zoology. For the use of Schools and Young Persons. Comprising the Principles of Classification interspersed with amusing and instructive original Accounts of the most remarkable Animals. By Mrs. R. LEE. New Edition, revised and enlarged, with numerous additional Woodcuts. Fcp. 8vo. 7s. 6d. cloth.

LEE.—TAXIDERMY;

Or, the Art of Collecting, Preparing, and Mounting Objects of Natural History. For the use of Museums and Travellers. By Mrs. R. LEE. New Edition, improved; with an account of a Visit to Walton Hall, and Mr. Waterton's Method of Preserving Animals. Fcp. 8vo. with Woodcuts, 7s. cloth.

L. E. L.—THE POETICAL WORKS OF LETITIA ELIZABETH

LANDON; comprising the IMPROVISATRICE, the VENETIAN BRACELET, the GOLDEN VIOLET, the TROUBADOUR, and other Poetical Remains. New Edition, uniform with Moore's *Songs, Ballads, and Sacred Songs*; with 2 Vignettes by Richard Doyle. 2 vols. 16mo. 10s. cloth; morocco, 21s.

₊ Also, an Edition, in 4 vols. fcp. 8vo. with Illustrations by Howard, &c. 28s. cloth; or £2. 4s. bound in morocco.

LETTERS ON HAPPINESS, ADDRESSED TO A FRIEND.

By the Authoress of " Letters to My Unknown Friends," " Twelve Years Ago, a Tale," " Some Passages from Modern History," and " Discipline." Fcp. 8vo. 6s. cloth.

LETTERS TO MY UNKNOWN FRIENDS.

By A LADY, Authoress of " Letters on Happiness," " Twelve Years Ago," " Discipline," and " Some Passages from Modern History." 2d Edition. Fcp. 8vo. 6s. cloth.

LINDLEY.—INTRODUCTION TO BOTANY.

By J. LINDLEY, Ph.D. F.R.S. L.S. &c. Professor of Botany in University College, London. New Edition, with Corrections and copious Additions. 2 vols. 8vo. with Six Plates and numerous Woodcuts, 24s. cloth.

LINWOOD.—ANTHOLOGIA OXONIENSIS,

Sive Florilegium e lusibus poeticis diversorum Oxoniensium Græcis et Latinis decerptum. Curante GULIELMO LINWOOD, M.A. Ædis Christi Alumno. 8vo. 14s. cloth.

LORIMER.—LETTERS TO A YOUNG MASTER MARINER

On some Subjects connected with his Calling. By the late CHARLES LORIMER. New Edition. Fcp. 8vo. 5s. 6d. cloth.

LOUDON.—THE AMATEUR GARDENER'S CALENDAR:

Being a Monthly Guide as to what should be avoided, as well as what should be done, in a Garden in each Month : with plain Rules *how to do* what is requisite; Directions for Laying Out and Planting Kitchen and Flower Gardens, Pleasure Grounds, and Shrubberies; and a short Account, in each Month, of the Quadrupeds, Birds, and Insects then most injurious to Gardens. By Mrs. LOUDON. 16mo with Wood Engravings, 7s. 6d. cloth.

LOUDON.—THE LADY'S COUNTRY COMPANION;

Or, How to Enjoy a Country Life Rationally. By Mrs. LOUDON, Author of "Gardening for Ladies," &c. New Edition. Fcp. 8vo. with Plate and Wood Engravings, 7s. 6d. cloth.

LOUDON'S SELF-INSTRUCTION FOR YOUNG GARDENERS,

Foresters, Bailiffs, Land Stewards, and Farmers; in Arithmetic, Book-keeping, Geometry, Mensuration, Practical Trigonometry, Mechanics, Land-Surveying, Levelling, Planning and Mapping, Architectural Drawing, and Isometrical Projection and Perspective; with Examples shewing their applications to Horticulture and Agricultural Purposes. With a Portrait of Mr. Loudon, and a Memoir by Mrs. Loudon. 8vo. with Woodcuts, 7s. 6d. cloth.

LOUDON'S ENCYCLOPÆDIA OF GARDENING;

Comprising the Theory and Practice of Horticulture, Floriculture, Arboriculture, and Landscape Gardening : including all the latest improvements ; a General History of Gardening in all Countries ; and a Statistical View of its Present State : with Suggestions for its Future Progress in the British Isles. Illustrated with many hundred Engravings on Wood by Branston. An entirely New Edition (1850), corrected throughout and considerably improved by Mrs. LOUDON. In One large Volume, 8vo. 50s. cloth.

LOUDON'S ENCYCLOPÆDIA OF TREES AND SHRUBS:

being the *Arboretum et Fruticetum Britannicum* abridged : containing the Hardy Trees and Shrubs of Great Britain, Native and Foreign, Scientifically and Popularly Described ; with their Propagation, Culture, and Uses in the Arts ; and with Engravings of nearly all the Species. Adapted for the use of Nurserymen, Gardeners, and Foresters. 8vo. with 2,000 Engravings on Wood, £2. 10s. cloth.

LOUDON'S ENCYCLOPÆDIA OF AGRICULTURE:

Comprising the Theory and Practice of the Valuation, Transfer, Laying-out, Improvement, and Management of Landed Property, and of the Cultivation and Economy of the Animal and Vegetable productions of Agriculture : including all the latest Improvements, a general History of Agriculture in all Countries, a Statistical View of its present State, with Suggestions for its future progress in the British Isles. New Edition ; with upwards of 1,100 Engravings on Wood. In One large Volume, 8vo. £2. 10s. cloth.

LOUDON'S ENCYCLOPÆDIA OF PLANTS:

Including all the Plants which are now found in, or have been introduced into, Great Britain ; giving their Natural History, accompanied by such descriptions, engraved figures, and elementary details, as may enable a beginner, who is a mere English reader, to discover the name of every Plant which he may find in flower, and acquire all the information respecting it which is useful and interesting. The Specific Characters by an Eminent Botanist; the Drawings by J. D. C. Sowerby. New Edition with Supplement, and new General Index. 8vo. with nearly 10,000 Wood Engravings, £3. 13s. 6d. cloth.

LOUDON'S ENCYCLOPÆDIA OF COTTAGE, FARM, AND

VILLA ARCHITECTURE and FURNITURE ; containing numerous Designs, from the Villa to the Cottage and the Farm, including Farm Houses, Farmeries, and other Agricultural Buildings ; Country Inns, Public Houses, and Parochial Schools; with the requisite Fittings-up, Fixtures, and Furniture, and appropriate Offices, Gardens, and Garden Scenery : each Design accompanied by Analytical and Critical Remarks. New Edition, edited by Mrs. LOUDON. 8vo. with more than 2,000 Engravings on Wood, £3. 3s. cloth.

D

LOUDON'S HORTUS BRITANNICUS;

Or, Catalogue of all the Plants indigenous to, cultivated in, or introduced into Britain. An entirely New Edition (1850), corrected throughout : with a Supplement, including all the New Plants down to March, 1850 ; and a New General Index to the whole Work. Edited by Mrs. LOUDON ; assisted by W. H. BAXTER, Esq., and DAVID WOOSTER. 8vo. 31s. 6d. cloth.

SUPPLEMENT TO LOUDON'S HORTUS BRITANNICUS;

Including all the Plants introduced into Britain, all the newly-discovered British Species, and all the kinds originated in British Gardens, up to March 1850. With a New General Index to the whole Work, including the Supplement. Prepared by W. H. BAXTER, Esq. ; assisted by D. WOOSTER, under the direction of Mrs. LOUDON. 8vo. 14s. cloth.

LOW.—ELEMENTS OF PRACTICAL AGRICULTURE;

Comprehending the Cultivation of Plants, the Husbandry of the Domestic Animals, and the Economy of the Farm. By D. Low, Esq. F.R.S.E. New Edition, with Alterations and Additions, and an entirely new set of above 200 Woodcuts. 8vo. 21s. cloth.

LOW.—ON LANDED PROPERTY,

And the ECONOMY of ESTATES ; comprehending the Relation of Landlord and Tenant, and the Principles and Forms of Leases ; Farm-Buildings, Enclosures, Drains, Embankments, and other Rural Works ; Minerals ; and Woods. By DAVID LOW, Esq. F.R.S.E. 8vo. with numerous Wood Engravings, 21s. cloth.

MACAULAY.—THE HISTORY OF ENGLAND FROM THE

ACCESSION OF JAMES II. By THOMAS BABINGTON MACAULAY. New Edition. Vols. 1. and 11. 8vo. 32s. cloth.

MACAULAY.—CRITICAL AND HISTORICAL ESSAYS CON-

TRIBUTED to The EDINBURGH REVIEW. By THOMAS BABINGTON MACAULAY. New Edition, complete in One Volume ; with Portrait by E. U. Eddis, engraved in line by W. Greatbach, and Vignette. Square crown 8vo. 21s. cloth ; 30s. calf extra by Hayday.—Or in 3 vols. 8vo. 36s. cloth.

MACAULAY.—LAYS OF ANCIENT ROME.

With " Ivry" and " The Armada." By THOMAS BABINGTON MACAULAY. New Edition. 16mo. 4s. 6d. cloth ; or 10s. 6d. bound in morocco by Hayday.

MR. MACAULAY'S LAYS OF ANCIENT ROME.

With numerous Illustrations, Original and from the Antique, drawn on Wood by George Scharf, Jun. and engraved by Samuel Williams. New Edition. Fcp. 4to. 21s. boards ; or 42s. bound in morocco by Hayday.

MACDONALD.—VILLA VEROCCHIO;

Or, the YOUTH of LEONARDO DA VINCI : a Tale. By the late DIANA LOUISA MACDONALD. Fcp. 8vo. 6s. cloth.

" An exceedingly agreeable volume, full of feeling and interest." EXAMINER.
" In this most pleasing of historiettes we have an episode from the life of one who carved out for himself the highest place among the great master minds of his age.... The scenery of the locality [the Val d'Arno and Firenze], the manners of the day, the characters of the great men of that age, and the affections of the best of the world's race, which go far to console us for our sad lot here below, are sketched with a facility, a correctness, and a delicacy, that fail not of carrying the reader, without a moment's stop, from the first to the last sentence of this little volume." BRITANNIA.

MACKINTOSH.—SIR JAMES MACKINTOSH'S MISCELLA-

NEOUS WORKS ; including his Contributions to The EDINBURGH REVIEW. A New Edition (1851), complete in One Volume ; with Portrait engraved in line by W. Greatbach, and Vignette. Square crown 8vo. 21s. cloth ; or 30s. calf extra by Hayday.

M'CULLOCH.—A DICTIONARY, PRACTICAL, THEORETI-
CAL, AND HISTORICAL, OF COMMERCE AND COMMERCIAL NAVIGATION. Illustrated with Maps and Plans. By J. R. M'CULLOCH, Esq. New Edition, (1850), corrected,
enlarged, and improved; with a Supplement. 8vo. 50s. cloth; or 55s. half-bound in russia.

**⁎⁎⁎ The SUPPLEMENT to the last Edition, published in 1849, may be had separately, price
4s. 6d. sewed.

M'CULLOCH.—A DICTIONARY, GEOGRAPHICAL, STATIS-
TICAL, AND HISTORICAL, of the various Countries, Places, and Principal Natural Objects
in the WORLD. By J. R. M'CULLOCH, Esq. Illustrated with 6 large Maps. New Edition
(1850-1851), corrected, and in part re-written; with a Supplement. 2 thick vols. 8vo. 63s.
cloth.

M'CULLOCH.—AN ACCOUNT, DESCRIPTIVE AND STATIS-
TICAL, of the BRITISH EMPIRE; exhibiting its Extent, Physical Capacities, Population,
Industry, and Civil and Religious Institutions. By J. R. M'CULLOCH, Esq. New Edition,
corrected, enlarged, and greatly improved. 2 thick vols. 8vo. 42s. cloth.

M'CULLOCH. — A TREATISE ON THE PRINCIPLES AND
PRACTICAL INFLUENCE of TAXATION and the FUNDING SYSTEM. By J. R.
M'CULLOCH, Esq. 8vo. 10s. cloth.

MAITLAND.—THE CHURCH IN THE CATACOMBS:
A Description of the Primitive Church of Rome. Illustrated by its Sepulchral Remains. By
CHARLES MAITLAND. New Edition, corrected. 8vo. with numerous Wood Engravings,
14s. cloth.

MARCET.—CONVERSATIONS ON CHEMISTRY;
In which the Elements of that Science are familiarly Explained and Illustrated by Experiments.
By JANE MARCET. New Edition, enlarged and improved. 2 vols. fcp. 8vo. 14s. cloth.

MARCET.—CONVERSATIONS ON NATURAL PHILOSOPHY;
In which the Elements of that Science are familiarly explained. By JANE MARCET. New
Edition, enlarged and corrected. Fcp. 8vo. with 23 Plates, 10s. 6d. cloth.

MARCET.—CONVERSATIONS ON POLITICAL ECONOMY;
In which the Elements of that Science are familiarly explained. By JANE MARCET. New
Edition revised and enlarged. Fcp. 8vo. 7s. 6d. cloth.

MARCET. — CONVERSATIONS ON VEGETABLE PHYSIO-
LOGY; comprehending the Elements of Botany, with their application to Agriculture.
By JANE MARCET. New Edition. Fcp. 8vo. with 4 Plates, 9s. cloth.

MARCET.—CONVERSATIONS ON LAND AND WATER.
By JANE MARCET. New Edition, revised and corrected. With a coloured Map, shewing
the comparative altitude of Mountains. Fcp. 8vo. 5s. 6d. cloth.

"This work consists of desultory Conversations with a family of children from six to ten years of age, in which
the author has endeavoured to mingle information with amusement, and to teach the youthful student of geography
that there are other matters connected with land and water quite as interesting as the names and situations of the
different parts of the earth. Two new Conversations have been added to this edition, containing the ' Adventures of
a Drop of Water.' "   PREFACE.

MARRYAT.—MASTERMAN READY;
Or, the Wreck of the Pacific. Written for Young People. By Captain F. MARRYAT, C.B.
Author of " Peter Simple," &c. 3 vols. fcp. 8vo. with Wood Engravings, 22s. 6d. cloth.

MARRYAT.—THE MISSION;
Or, Scenes in Africa. Written for Young People. By Captain F. MARRYAT, C.B. Author of
" Masterman Ready," &c. 2 vols. fcp. 8vo. 12s. cloth.

MARRYAT. — THE PRIVATEER'S-MAN ONE HUNDRED
YEARS AGO. By Captain F. MARRYAT, C.B. Author of "Masterman Ready," &c. 2 vols.
fcp. 8vo. 12s. cloth.

MARRYAT.—THE SETTLERS IN CANADA.
Written for Young People. By Captain F. MARRYAT, C.B. Author of "Masterman Ready,"
&c. New Edition. Fcp. 8vo. with 2 Illustrations, 7s. 6d. cloth.

MAUNDER.—THE BIOGRAPHICAL TREASURY;
Consisting of Memoirs, Sketches, and brief Notices of above 12,000 Eminent Persons of all Age.
and Nations, from the Earliest Period of History; forming a new and complete Dictionary
of Universal Biography. By SAMUEL MAUNDER. A New and carefully-revised Edition
(1851); corrected throughout, and brought down to the Present Time, by the introduction of
numerous additional Lives. Fcp. 8vo. 10s. cloth; bound in roan. 12s.

MAUNDER.—THE TREASURY OF HISTORY;
Comprising a General Introductory Outline of Universal History, Ancient and Modern, and a
Series of separate Histories of every principal Nation that exists; their Rise, Progress, and
Present Condition, the Moral and Social Character of their respective inhabitants, their
Religion, Manners, and Customs, &c. By SAMUEL MAUNDER. New Edition. Fcp. 8vo. 10s.
cloth; bound in roan, 12s.

MAUNDER.—THE SCIENTIFIC & LITERARY TREASURY;
A new and popular Encyclopædia of Science and the Belles-Lettres; including all Branches of
Science, and every Subject connected with Literature and Art. The whole written in a familiar
style, adapted to the comprehension of all persons desirous of acquiring information on the
subjects comprised in the work, and also adapted for a Manual of convenient Reference to the
more instructed. By S. MAUNDER. New Edition. Fcp. 8vo. 10s. cloth; bound in roan, 12s.

MAUNDER.—THE TREASURY OF NATURAL HISTORY;
Or, a Popular Dictionary of Animated Nature: in which the Zoological Characteristics that
distinguish the different Classes, Genera, and Species are combined with a variety of interesting Information illustrative of the Habits, Instincts and General Economy of the Animal
Kingdom. To which are added, a Syllabus of Practical Taxidermy, and a Glossarial
Appendix. Embellished with 900 accurate Engravings on Wood, from Drawings made
expressly for this work. By SAMUEL MAUNDER. New Edition. Fcp. 8vo. 10s. cloth,
bound in roan, 12s.

MAUNDER.—THE TREASURY OF KNOWLEDGE,
And LIBRARY of REFERENCE. Comprising an English Grammar; Tables of English
Verbal Distinctions; Proverbs, Terms, and Phrases, in Latin, Spanish, French, and Italian,
translated; New and Enlarged English Dictionary; Directions for Pronunciation; New
Universal Gazetteer; Tables of Population and Statistics; List of Cities, Boroughs, and
Market Towns in the United Kingdom; Regulations of the General Post Office; List of Foreign
Animal, Vegetable, and Mineral Productions; Compendious Classical Dictionary; Scripture
Proper Names accented, and Christian Names of Men and Women: with Latin Maxims
translated; List of Abbreviations; Chronology and History; compendious Law Dictionary;
Abstract of Tax Acts; Interest and other Tables; Forms of Epistolary Address; Tables of
Precedency; Synopsis of the British Peerage; and Tables of Number, Money, Weights, and
Measures By SAMUEL MAUNDER. 18th Edition, revised throughout, and greatly enlarged.
Fcp. 8vo. 10s. cloth; bound in roan, 12s.

MEMOIRS OF THE GEOLOGICAL SURVEY OF GREAT
BRITAIN, and of the Museum of Economic Geology in London. Published by order of the
Lords Commissioners of Her Majesty's Treasury. Royal 8vo. with Woodcuts and 9 large
Plates (seven coloured), 21s. cloth; and Vol. II. in Two thick Parts, with 63 Plates (three
coloured), and numerous Woodcuts, 42s. cloth, or, separately, 21s. each Part.——Also,
BRITISH ORGANIC REMAINS; consisting of Plates of Figures engraved on Steel, with
descriptive Letterpress, and forming a portion of the Memoirs of the Geological Survey.
Decades I. to III. royal 8vo. 2s. 6d. each; or, royal 4to. 4s. 6d. each, sewed.

MERIVALE. — A HISTORY OF THE ROMANS UNDER
THE EMPIRE. By the Rev. CHARLES MERIVALE, late Fellow and Tutor of St. John's College, Cambridge. Vols. I. and II. 8vo. 28s.

JAMES MONTGOMERY'S POETICAL WORKS.
With some additional Poems, and the Author's Autobiographical Prefaces. A New Edition, complete in One Volume, uniform with Southey's "The Doctor &c." and "Commonplace Book;" with Portrait and Vignette. Square crown 8vo. 10s. 6d. cloth; morocco, 21s.—Or, in 4 vols. fcp. 8vo. with Portrait, and Seven other Plates, 20s. cloth; morocco, 35s.

MOORE.—HEALTH, DISEASE, AND REMEDY,
Familiarly and practically considered in a few of their relations to the Blood. By GEORGE MOORE, M.D. Member of the Royal College of Physicians. Post 8vo. 7s. 6d. cloth.

MOORE.—MAN AND HIS MOTIVES.
By GEORGE MOORE, M.D., Member of the Royal College of Physicians. New Edition. Post 8vo. 8s. cloth.

MOORE.—THE POWER OF THE SOUL OVER THE BODY,
Considered in relation to Health and Morals. By GEORGE MOORE, M.D. Member of the Royal College of Physicians. New Edition. Post 8vo. 7s. 6d. cloth.

MOORE.—THE USE OF THE BODY IN RELATION TO THE
MIND. By GEORGE MOORE, M.D. Member of the Royal College of Physicians. New Edition. Post 8vo. 9s. cloth.

THOMAS MOORE'S POETICAL WORKS;
Containing the Author's recent Introduction and Notes. Complete in One Volume, uniform with Lord Byron's and Southey's Poems. With a Portrait by George Richmond, engraved in line, and a View of Sloperton Cottage. Medium 8vo. 21s. cloth; morocco by Hayday, 42s. —Or, in 10 vols. fcp. 8vo. with Portrait, and 19 Plates, £2. 10s. cloth; morocco, £4. 10s.

MOORE.—SONGS, BALLADS, AND SACRED SONGS.
By THOMAS MOORE, Author of "Lalla Rookh," "Irish Melodies," &c. First collected Edition, uniform with the smaller Edition of Mr. Macaulay's *Lays of Ancient Rome;* with Vignette by R. Doyle. 16mo. 5s. cloth; 12s. 6d. smooth morocco, by Hayday.

MOORE'S IRISH MELODIES.
New Edition, uniform with the smaller Edition of Mr. Macaulay's *Lays of Ancient Rome.* With the Autobiographical Preface from the Collective Edition of Mr. Moore's Poetical Works, and a Vignette Title by D. Maclise, R.A. 16mo. 5s. cloth; 12s. 6d. smooth morocco, by Hayday.—Or, in fcp. 8vo. 10s. cloth; bound in morocco, 13s. 6d.

MOORE'S IRISH MELODIES.
Illustrated by D. MACLISE, R.A. Imperial 8vo. with 161 Steel Plates, £3. 3s. boards; or £4. 14s. 6d. bound in morocco by Hayday. Proof Impressions (only 200 copies printed, of which a very few now remain), £6. 6s boards.

MOORE'S LALLA ROOKH: AN ORIENTAL ROMANCE.
New Edition, uniform with the smaller Edition of Mr. Macaulay's *Lays of Ancient Rome.* With the Autobiographical Preface from the Collective Edition of Mr. Moore's Poetical Works, and a Vignette Title by D. Maclise, R.A. 16mo. 5s. cloth; 12s. 6d. smooth morocco, by Hayday.—Or, in fcp. 8vo. with Four Engravings from Paintings by Westall, 10s. 6d. cloth; bound in morocco, 14s.

MOORE'S LALLA ROOKH: AN ORIENTAL ROMANCE.
With 13 Plates from Designs by Corbould, Meadows, and Stephanoff, engraved under the superintendence of Mr. Charles Heath. Royal 8vo. 21s. cloth; morocco, 35s; or, with India Proof Plates, 42s. cloth.

MORELL.—THE PHILOSOPHY OF RELIGION.
By J. D. Morell, M.A. Author of an Historical and Critical *View of the Speculative Philosophy of Europe in the Nineteenth Century.* 8vo. 12s. cloth.

MOSELEY.—THE MECHANICAL PRINCIPLES OF ENGI-
NEERING AND ARCHITECTURE. By the Rev. H. MOSELEY, M.A. F.R.S., Professor of
Natural Philosophy and Astronomy in King's College, London. 8vo. with Woodcuts and
Diagrams, 24s. cloth.

MOSELEY.—ILLUSTRATIONS OF PRACTICAL MECHANICS.
By the Rev. H. MOSELEY, M.A., Professor of Natural Philosophy and Astronomy in King's
College, London. New Edition. Fcp. 8vo. with numerous Woodcuts, 8s. cloth.

MOSHEIM'S ECCLESIASTICAL HISTORY,
Ancient and Modern. Translated, with copious Notes, by JAMES MURDOCK, D.D. New
Edition, revised, and continued to the Present Time, by the Rev. HENRY SOAMES, M.A.
4 vols. 8vo. 48s. cloth.

MOUNT SAINT LAWRENCE.
By the Author of "Mary the Star of the Sea." 2 vols. post 8vo. 12s. cloth.

MURE.—A CRITICAL HISTORY OF THE LANGUAGE AND
LITERATURE OF ANCIENT GREECE. By WILLIAM MURE, M.P., of Caldwell. 3 vols.
8vo. 36s. cloth.

MURRAY.—ENCYCLOPÆDIA OF GEOGRAPHY;
Comprising a complete Description of the Earth: exhibiting its Relation to the Heavenly
Bodies, its Physical Structure, the Natural History of each Country, and the Industry, Com-
merce, Political Institutions, and Civil and Social State of all Nations. By HUGH MURRAY,
F.R.S.E.: assisted by other Writers of eminence. Second Edition. 8vo. with 82 Maps, and
upwards of 1,000 other Woodcuts, £3, cloth.

NEALE.—THE EARTHLY RESTING PLACES OF THE JUST.
By the Rev. ERSKINE NEALE, M.A., Rector of Kirton, Suffolk; Author of "The Closing
Scene," &c. With Wood Engravings. Fcp. 8vo. 7s. cloth.

NEALE.—THE CLOSING SCENE;
Or, Christianity and Infidelity contrasted in the Last Hours of Remarkable Persons. By the
Rev. ERSKINE NEALE, M.A., Rector of Kirton, Suffolk; Author of "The Earthly Resting-
places of the Just," &c. New Editions of the First and Second Series. 2 vols. fcp. 8vo. 12s.
cloth; or separately, 6s. each.

NEWMAN.—DISCOURSES ADDRESSED TO MIXED CON-
GREGATIONS. By JOHN HENRY NEWMAN, Priest of the Oratory of St. Philip Neri.
Second Edition. 8vo. 12s. cloth.

OWEN JONES.—WINGED THOUGHTS:
A Series of Poems. By MARY ANNE BACON. With Illustrations of Birds, designed by
E. L. Bateman, and executed in Illuminated Printing by Owen Jones. Uniform with *Flowers
and their Kindred Thoughts* and *Fruits from the Garden and the Field.* Imperial 8vo. 31s. 6d.
elegantly bound in calf.

OWEN JONES. — FLOWERS AND THEIR KINDRED
THOUGHTS: A Series of Stanzas. By MARY ANNE BACON, Authoress of "Winged
Thoughts." With beautiful Illustrations of Flowers, designed and printed in Colours by
Owen Jones. Uniform with *Fruits from the Garden and the Field.* Imperial 8vo. 31s. 6d.
elegantly bound in calf.

OWEN JONES.—FRUITS FROM THE GARDEN AND THE
FIELD. A Series of Stanzas. By MARY ANNE BACON, Authoress of "Winged Thoughts."
With beautiful Illustrations of Fruit, designed and printed in Colours by Owen Jones. Uniform with *Flowers and their Kindred Thoughts.* Imperial 8vo. 31s. 6d. elegantly bound in
calf.

OWEN JONES'S ILLUMINATED EDITION OF GRAY'S
ELEGY. GRAY'S ELEGY, WRITTEN IN A COUNTRY CHURCHYARD. Illuminated, in the
Missal Style, by OWEN JONES, Architect. Imperial 8vo. 31s. 6d. elegantly bound.

OWEN JONES'S ILLUMINATED EDITION OF THE SERMON
ON THE MOUNT. THE SERMON ON THE MOUNT. Printed in Gold and Colours, in the Missal Style; with Ornamental Borders by Owen Jones, and an Illuminated Frontispiece by W. Boxall. New Edition. Square fcp. 8vo. in rich silk covers, 21s. ; or bound in morocco by Hayday, 25s.

OWEN JONES'S ILLUMINATED EDITION OF THE MAR-
RIAGE SERVICE. THE FORM OF SOLEMNISATION OF MATRIMONY. From *The Book of Common Prayer*. Illuminated, in the Missal Style, by Owen Jones. Square 18mo. 21s. elegantly bound in white calf.

OWEN JONES'S ILLUMINATED EDITION OF THE
PREACHER. The Words of the Preacher, Son of David, King of Jerusalem. From the Holy Scriptures. Being the Twelve Chapters of the Book of Ecclesiastes, elegantly Illuminated, in the Missal Style, by Owen Jones. Imperial 8vo. in very massive carved covers, 42s. ; or, handsomely bound in calf, 31s. 6d.

OWEN JONES'S ILLUMINATED EDITION OF SOLOMON'S
SONG. THE SONG OF SONGS, WHICH IS SOLOMON'S. From the Holy Scriptures. Being the Six Chapters of the Book of the Song of Solomon, richly Illuminated, in the Missal Style, by Owen Jones. Elegantly bound in relievo leather. Imperial 16mo. 21s.

OWEN JONES'S TRANSLATION OF D'AGINCOURT'S HIS-
TORY OF ART. THE HISTORY OF ART, BY ITS MONUMENTS, from its Decline in the Fourth Century to its Restoration in the Sixteenth. Translated from the French of SEROUX D'AGINCOURT, by Owen Jones, Architect. In 3,335 Subjects, engraved on 328 Plates. Vol. I. Architecture, 73 Plates ; Vol. II. Sculpture, 51 Plates ; Vol. III. Painting, 204 Plates. 3 vols. royal folio, £5. 5s. sewed.

OWEN. — LECTURES ON THE COMPARATIVE ANATOMY
and PHYSIOLOGY of the INVERTEBRATE ANIMALS, delivered at the Royal College of Surgeons in 1843. By Richard Owen, F.R.S. Hunterian Professor to the College. New Edition, corrected. 8vo. with very numerous Wood Engravings. [*Nearly ready.*

OWEN.—LECTURES ON THE COMPARATIVE ANATOMY
and PHYSIOLOGY of the VERTEBRATE ANIMALS, delivered at the Royal College of Surgeons in 1844 and 1846. By Richard Owen, F.R.S. Hunterian Professor to the College. In 2 vols. Vol. I. 8vo. with numerous Woodcuts, 14s. cloth.

PALEY'S EVIDENCES OF CHRISTIANITY:
And Horæ Paulinæ. A New Edition, with Notes, an Analysis, and a Selection of Papers from the Senate-House and College Examination Papers. Designed for the Use of Students in the University. By Robert Potts, M.A. Trinity College, Cambridge. 8vo. 10s. 6d. cloth.

PASCAL'S ENTIRE WORKS, TRANSLATED BY PEARCE.
The COMPLETE WORKS of BLAISE PASCAL: With M. Villemain's Essay on Pascal considered as a Writer and Moralist prefixed to the *Provincial Letters*; and the *Miscellaneous Writings, Thoughts on Religion*, and *Evidences of Christianity* re-arranged, with large Additions, from the French Edition of Mons. P. Faugère. Newly Translated from the French, with Memoir, Introductions to the various Works, Editorial Notes, and Appendices, by George Pearce, Esq. 3 vols. post 8vo. with Portrait, 25s. 6d. cloth.

* *The Three Volumes may be had separately, as follows:—*

Vol. I.— PASCAL'S PROVINCIAL LETTERS: with M. Villemain's Essay on Pascal prefixed, and a new Memoir. Post 8vo. Portrait, 8s. 6d.

Vol. II.— PASCAL'S THOUGHTS on RELIGION and EVIDENCES of CHRISTIANITY, with Additions, from Original MSS. : from M. Faugère's Edition. Post 8vo. 8s. 6d.

Vol. III.— PASCAL'S MISCELLANEOUS WRITINGS, Correspondence, Detached Thoughts, &c. : from M. Faugère's Edition. Post 8vo. 8s. 6d.

PEREIRA.—A TREATISE ON FOOD AND DIET:

With Observations on the Dietetical Regimen suited for Disordered States of the Digestive Organs; and an Account of the Dietaries of some of the principal Metropolitan and other Establishments for Paupers, Lunatics, Criminals, Children, the Sick, &c. By JON. PEREIRA, M.D. F.R.S. & L.S. Author of " Elements of Materia Medica." 8vo. 16s. cloth.

PESCHEL.—ELEMENTS OF PHYSICS.

By C. F. PESCHEL, Principal of the Royal Military College, Dresden. Translated from the German, with Notes, by E. WEST. 3 vols. fcp. 8vo. with Diagrams and Woodcuts, 21s. cloth.

Separately { Part 1. The Physics of Ponderable Bodies. Fcp. 8vo. 7s. 6d. cloth.
Part 2. Imponderable Bodies (Light, Heat, Magnetism, Electricity, and Electro-Dynamics). 2 vols. fcp. 8vo 13s. 6d. cloth.

PHILLIPS.—AN ELEMENTARY INTRODUCTION TO MINE-

RALOGY; comprising a Notice of the Characters, Properties, and Chemical Constitution of Minerals: with Accounts of the Places and Circumstances in which they are found. By WILLIAM PHILLIPS, F.L.S.M.G.S. &c. A New Edition, corrected, enlarged, and improved, by H. J. BROOKE, F.R.S.; and W. H. MILLER, M.A., F.R.S., Professor of Mineralogy in the University of Cambridge. Post 8vo. with numerous Wood Engravings. [In the press.

PHILLIPS.—FIGURES AND DESCRIPTIONS OF THE

PALÆOZOIC FOSSILS of CORNWALL, DEVON, and WEST SOMERSET; observed in the course of the Ordnance Geological Survey of that District. By JOHN PHILLIPS, F.R.S. F.G.S. &c. Published by Order of the Lords Commissioners of H.M. Treasury. 8vo. with 60 Plates, comprising very numerous figures, 9s cloth.

PORTLOCK.—REPORT ON THE GEOLOGY OF THE COUNTY

of LONDONDERRY, and of Parts of Tyrone and Fermanagh, examined and described under the Authority of the Master-General and Board of Ordnance. By J. E. PORTLOCK, F.R.S &c. 8vo. with 48 Plates, 24s. cloth.

POWER.—SKETCHES IN NEW ZEALAND,

with Pen and Pencil. By W. TYRONE POWER, D.A.C.G. From a Journal kept in that Country, from July 1846 to June 1848. With 8 Plates and 2 Woodcuts, from Drawings made on the spot. Post 8vo. 12s. cloth.

PULMAN.—THE VADE-MECUM OF FLY-FISHING FOR

TROUT: being a complete Practical Treatise on that Branch of the Art of Angling; with plain and copious Instructions for the Manufacture of Artificial Flies. By G. P. R. PULMAN, Author of "The Book of the Axe." Third Edition, re-written and greatly enlarged; with several Woodcuts. Fcp. 8vo. 6s. cloth.

PYCROFT.—A COURSE OF ENGLISH READING,

Adapted to every Taste and Capacity: with Literary Anecdotes. By the Rev. JAMES PYCROFT, B.A. Author of "The Collegian's Guide, &c." New Edition. Fcp. 8vo. 5s. cloth.

DR. REECE'S MEDICAL GUIDE;

For the Use of the Clergy, Heads of Families, Schools, and Junior Medical Practitioners; comprising a complete Modern Dispensatory, and a Practical Treatise on the distinguishing Symptoms, Causes, Prevention, Cure, and Palliation of the Diseases incident to the Human Frame. With the latest Discoveries in the different departments of the Healing Art, Materia Medica, &c. Seventeenth Edition (1850), with considerable Additions; revised and corrected by the Author's Son, Dr. HENRY REECE, M.R.C.S &c. 8vo. 12s. cloth.

RICH.—THE ILLUSTRATED COMPANION TO THE LATIN

DICTIONARY AND GREEK LEXICON: forming a Glossary of all the Words representing Visible Objects connected with the Arts, Manufactures, and Every-day Life of the Ancients. With Representations of nearly Two Thousand Objects from the Antique. By ANTHONY RICH, Jun. B.A. late of Caius College, Cambridge. Post 8vo. with about 2,000 Woodcuts, 21s. cloth.

RICHARDSON.—NARRATIVE OF AN OVERLAND JOURNEY
in SEARCH of the DISCOVERY SHIPS under SIR JOHN FRANKLIN, in the YEARS 1847,
1848, and 1849. By Sir JOHN RICHARDSON, M.D., F.R.S., &c., Inspector of Hospitals. Published by Authority of the Admiralty. 2 vols. 8vo. with Maps and Plans. [*In the press.*

RIDDLE.—A COPIOUS AND CRITICAL LATIN-ENGLISH
LEXICON, founded on the German-Latin Dictionaries of Dr. William Freund. By the Rev.
J. E. RIDDLE, M.A. of St. Edmund's Hall, Oxford. Uniform with *Yonge's English Greek
Lexicon.* New Edition. Post 4to. £2. 10s. cloth.

RIDDLE.—A COMPLETE LATIN-ENGLISH AND ENGLISH-
LATIN DICTIONARY, for the use of Colleges and Schools. By the Rev. J. E. RIDDLE, M.A.
of St. Edmund Hall, Oxford. New Edition, revised and corrected. 8vo. 31s. 6d. cloth.
Separately { The English-Latin Dictionary, 10s. 6d. cloth.
{ The Latin-English Dictionary, 21s. cloth.

RIDDLE. — A DIAMOND LATIN-ENGLISH DICTIONARY.
For the Waistcoat-pocket. A Guide to the Meaning, Quality, and right Accentuation of Latin
Classical Words. By the Rev. J. E. RIDDLE, M.A. New Edition. Royal 32mo. 4s. bound.

RIVERS.—THE ROSE AMATEUR'S GUIDE;
Containing ample Descriptions of all the fine leading varieties of Roses, regularly classed in
their respective Families; their History and mode of Culture. By T. RIVERS, Jun. New
Edition, corrected and improved. Fcp. 8vo. 6s. cloth.

ROBINSON'S LEXICON TO THE GREEK TESTAMENT.
A GREEK and ENGLISH LEXICON of the NEW TESTAMENT. By EDWARD ROBINSON,
D.D., LL.D., Professor of Biblical Literature in the Union Theological Seminary, New York;
Author of "Biblical Researches in Palestine," &c. A New Edition (1850), revised, and in
great part re-written. In One large Volume, 8vo. 18s. cloth.

ROGERS. — ESSAYS SELECTED FROM CONTRIBUTIONS
To the EDINBURGH REVIEW. By HENRY ROGERS. 2 vols. 8vo. 24s. cloth.

RONALDS.—THE FLY-FISHER'S ENTOMOLOGY.
Illustrated by coloured Representations of the Natural and Artificial Insect; and accompanied by a few Observations and Instructions relative to Trout and Grayling Fishing. By
ALFRED RONALDS. 4th Edition, corrected; with Twenty Copperplates. 8vo. 14s. cloth.

ROVINGS IN THE PACIFIC, FROM 1837 TO 1849;
With a GLANCE at CALIFORNIA. By A MERCHANT LONG RESIDENT AT TAHITI. With
Four Illustrations printed in colours. 2 vols. post 8vo. 21s. cloth.

ROWTON.—THE DEBATER ;
Being a Series of complete Debates, Outlines of Debates, and Questions for Discussion ; with
ample References to the best Sources of Information on each particular Topic. By FREDERIC
ROWTON, Author of "The Female Poets of Great Britain." New Edition. Fcp. 8vo. 6s. cloth.

SCHLEIDEN.—PRINCIPLES OF SCIENTIFIC BOTANY;
Or, Botany as an Inductive Science. By Dr. M. J. SCHLEIDEN, Extraordinary Professor of
Botany in the University of Jena. Translated by EDWIN LANKESTER, M.D. F.R.S. F.L.S.
Lecturer on Botany at the St. George's School of Medicine, London. 8vo. with Plates and
Woodcuts, 21s. cloth.

SCOFFERN.—THE MANUFACTURE OF SUGAR,
In the Colonies and at Home, chemically considered. By JOHN SCOFFERN, M.B. Lond. late
Professor of Chemistry at the Aldersgate College of Medicine. 8vo. with Illustrations (one
coloured) 10s. 6d. cloth.

SEAWARD.—SIR EDWARD SEAWARD'S NARRATIVE OF

HIS SHIPWRECK, and consequent Discovery of certain Islands in the Caribbean Sea: with a detail of many extraordinary and highly interesting Events in his Life, from 1733 to 1749, as written in his own Diary. Edited by Miss JANE PORTER. Third Edition, with a New Nautical and Geographical Introduction. 2 vols. post 8vo. 21s. cloth.

SEWELL.—AMY HERBERT.

By a LADY. Edited by the Rev. WILLIAM SEWELL, B.D. Fellow and Tutor of Exeter College, Oxford. New Edition. 2 vols. fcp. 8vo. 9s. cloth.

SEWELL.—THE EARL'S DAUGHTER.

By the Authoress of "Amy Herbert," "Gertrude," "Laneton Parsonage," "Margaret Percival," and "The Child's History of Rome." Edited by the Rev. WILLIAM SEWELL, D.B. Fellow and Tutor of Exeter College, Oxford. 2 vols. fcp. 8vo. 9s. cloth.

SEWELL.—GERTRUDE.

A Tale. By the Authoress of "Amy Herbert." Edited by the Rev. WILLIAM SEWELL, B.D. Fellow and Tutor of Exeter College, Oxford. New Edition. 2 vols. fcp. 8vo. 9s. cloth.

SEWELL.—LANETON PARSONAGE:

A Tale for Children, on the Practical Use of a portion of the Church Catechism. By the Authoress of "Amy Herbert." Edited by the Rev. W. SEWELL, B.D. Fellow and Tutor of Exeter College, Oxford. New Edition. 3 vols. fcp. 8vo. 16s. cloth.

SEWELL.—MARGARET PERCIVAL.

By the Authoress of "Amy Herbert." Edited by the Rev. W. SEWELL, B.D. Fellow and Tutor of Exeter College, Oxford. New Edition. 2 vols. fcp. 8vo. 12s. cloth.

SHAKSPEARE, BY BOWDLER.

THE FAMILY SHAKSPEARE; in which nothing is *added* to the Original Text; but those words and expressions are *omitted* which cannot with propriety be read aloud. By T BOWDLER, Esq. F.R.S. New Edition. 8vo. with 36 Engravings on Wood, from designs by Smirke, Howard, and other Artists, 21s. cloth; or, in 8 vols. 8vo. without Illustrations, £4. 14s. 6d. boards.

SHARP'S BRITISH GAZETTEER.

A NEW and COMPLETE BRITISH GAZETTEER, or TOPOGRAPHICAL DICTIONARY of the UNITED KINGDOM. Containing a Description of every Place, and the principal Objects of Note, founded upon the Ordnance Surveys, the best Local and other Authorities, and the most recent Official Documents connected with Population, Constituencies, Corporate and Ecclesiastical Affairs, Poor Laws, Education, Charitable Trusts, Railways, Trade, &c. By J. A. SHARP. In *Two* very large *Volumes*, 8vo. uniform with Johnston's *New General Gazetteer of the World*. [*In the press.*

SHORT WHIST:

Its Rise, Progress, and Laws; with Observations to make any one a Whist Player; containing also the Laws of Piquet, Cassino, Ecarté, Cribbage, Backgammon. By Major A * * * * *. New Edition. To which are added, Precepts for Tyros. By Mrs. B * * * *. Fcp. 8vo. 3s cloth.

SINCLAIR.—THE BUSINESS OF LIFE.

By CATHERINE SINCLAIR, Author of "The Journey of Life," "Modern Society," "Jane Bouverie," &c. 2 vols. fcap 8vo. 10s. cloth.

SINCLAIR.—THE JOURNEY OF LIFE.

By CATHERINE SINCLAIR, Author of "The Business of Life," "Modern Society," "Jane Bouverie," &c. New Edition, corrected and enlarged. Fcp. 8vo. 5s. cloth.

SIR ROGER DE COVERLEY.
From *The Spectator*. With Notes and Illustrations, by W. HENRY WILLS; and Twelve fine Wood Engravings, by John Thompson, from Designs by FREDERICK TAYLER. Crown 8vo. 15s. boards; or 27s. bound in morocco by Hayday.

THE SKETCHES:
Three Tales. By the Authors of "Amy Herbert," "The Old Man's Home," and "Hawkstone." New Edition. Fcp. 8vo. with 6 Plates, 8s. cloth.

SMEE.—ELEMENTS OF ELECTRO-METALLURGY.
By ALFRED SMEE, F.R S., Surgeon to the Bank of England. Third Edition, revised, corrected, and considerably enlarged; with Electrotypes and numerous Woodcuts. Post 8vo. 10s. 6d. cloth.

SMITH.—THE WORKS OF THE REV. SYDNEY SMITH:
Including his Contributions to The Edinburgh Review. New Edition, complete in One Volume; with Portrait by E. U. Eddis, engraved in line by W. Greatbach, and View of Combe Florey Rectory, Somerset. Square crown 8vo. 21s. cloth; 30s. calf extra, by Hayday: or in 3 vols. 8vo. with Portrait, 36s. cloth.

SMITH.—ELEMENTARY SKETCHES OF MORAL PHILO-
SOPHY, delivered at the Royal Institution in the Years 1804, 1805, and 1806. By the late Rev. SYDNEY SMITH, M.A. With an Introductory Letter to Mrs. Sydney Smith from the late Lord Jeffrey. Second Edition. 8vo. 12s. cloth.

SMITH.—SERMONS PREACHED AT ST. PAUL'S CATHE-
DRAL, the Foundling Hospital, and several Churches in London; together with others addressed to a Country Congregation. By the late Rev. SYDNEY SMITH, Canon Residentiary of St. Paul's Cathedral. 8vo. 12s. cloth.

SMITH.—THE DOCTRINE OF THE CHERUBIM:
Being an Inquiry, Critical, Exegetical, and Practical, into the Symbolical Character and Design of the Cherubic Figure of Holy Scripture. By GEORGE SMITH, F.A.S., &c. Post 8vo. 3s. cloth.

SMITH.—SACRED ANNALS;
Or, Researches into the History and Religion of Mankind, from the Creation of the World to the Death of Isaac: deduced from the Writings of Moses and other Inspired Authors, copiously illustrated and confirmed by the Ancient Records, Traditions, and Mythology of the Heathen World. By GEORGE SMITH, F.A.S. Crown 8vo. 10s. cloth.

SMITH.—THE HEBREW PEOPLE;
Or, the History and Religion of the Israelites, from the Origin of the Nation to the Time of Christ: deduced from the Writings of Moses and other Inspired Authors; and illustrated by copious References to the Ancient Records, Traditions, and Mythology of the Heathen World. By GEORGE SMITH, F.A.S. &c. Forming the Second Volume of *Sacred Annals*. Crown 8vo. In Two Parts, 12s. cloth.

SMITH.— THE RELIGION OF ANCIENT BRITAIN HISTORI-
CALLY CONSIDERED: or, a Succinct Account of the several Religious Systems which have obtained in this Island from the Earliest Times to the Norman Conquest: including an Investigation into the Early Progress of Error in the Christian Church, the Introduction of the Gospel into Britain, and the State of Religion in England till Popery had gained the ascendancy. By GEORGE SMITH, F.A.S. New Edition. 8vo. 7s. 6d. cloth.

SMITH.—PERILOUS TIMES;
Or, the Aggressions of Antichristian Error on Scriptural Christianity, considered in reference to the Dangers and Duties of Protestants. By GEORGE SMITH, F.A.S. Fcp. 8vo. 6s. cloth.

SMITH.—THE VOYAGE AND SHIPWRECK OF ST. PAUL:
with Dissertations on the Sources of the Writings of St. Luke, and the Ships and Navigation of the Antients. By JAMES SMITH, Esq. of Jordan Hill, F.R.S. 8vo. with Views, Charts, and Woodcuts, 14s. cloth.

SNOW.—VOYAGE OF THE PRINCE ALBERT IN SEARCH OF
SIR JOHN FRANKLIN: A Narrative of Every-day Life in the Arctic Seas. By W. PARKER SNOW. With a Chart, and 4 Illustrations printed in Colours. Post 8vo. 12s. cloth.

THE LIFE AND CORRESPONDENCE OF THE LATE
ROBERT SOUTHEY. Edited by his Son, the Rev. CHARLES CUTHBERT SOUTHEY M.A., Vicar of Ardleigh. With numerous Portraits, and Six Landscape Illustrations from Designs by William Westall, A.R.A. 6 vols. post 8vo. 63s. cloth.

₊ Each of the Six Volumes may be had separately, price 10s. 6d.

SOUTHEY'S COMMONPLACE BOOK—FOURTH SERIES.
FOURTH and last SERIES; being ORIGINAL MEMORANDA, and comprising Collections, Ideas, and Studies for Literary Compositions in general; Collections for a History of English Literature and Poetry; Characteristic English Anecdotes, and Fragments for *Espriella*; Collections for *The Doctor &c.*; Personal Observations and Recollections, with Fragments of Journals; Miscellaneous Anecdotes and Gleanings; Extracts, Facts, and Opinions relating to Political and Social Society; Texts for Sermons; Texts for Enforcement; and L'Envoy: forming a Single Volume complete in itself. Edited by Mr. Southey's Son-in-law, the Rev. J. W. WARTER, B.D. Square crown 8vo. 21s. cloth.

SOUTHEY'S COMMONPLACE BOOK—THIRD SERIES.
Being ANALYTICAL READINGS; and comprising *Analytical Readings*, with Illustrations and copious *Extracts*, of Works in English Civil History; English Ecclesiastical History; Anglo-Irish History; French History; French Literature; Miscellaneous Foreign Civil History; General Ecclesiastical History; Historical Memoirs; Ecclesiastical Biography; Miscellaneous Biography; Correspondence; Voyages and Travels; Topography; Natural History; Divinity; Literary History; Miscellaneous Literature; and Miscellanies. Forming a Single Volume complete in itself. Edited by Mr. Southey's Son-in-law, the Rev. J. W. WARTER, B.D. Square crown 8vo. 21s. cloth.

SOUTHEY'S COMMONPLACE BOOK—SECOND SERIES.
Comprising SPECIAL COLLECTIONS—viz. Ecclesiasticals, or Notes and Extracts on Theological Subjects (with Collections concerning Cromwell's Age); Spanish and Portuguese Literature; Middle Ages, &c.; Notes for the History of the Religious Orders; Orientalia, or Eastern and Mahommedan Collections; American Tribes; Incidental and Miscellaneous Illustrations; Physica, or Remarkable Facts in Natural History; and Curious Facts, quite Miscellaneous. Forming a single Volume complete in itself. Edited by Mr. Southey's Son-in-Law, the Rev. J. W. WARTER, B.D. Square crown 8vo. 18s. cloth.

SOUTHEY'S COMMONPLACE BOOK FIRST SERIES.
Comprising CHOICE PASSAGES, Moral, Religious, Political, Philosophical, Historical, Poetical, and Miscellaneous; and COLLECTIONS for the History of Manners and Literature in England. Forming a single Volume complete in itself. Edited by Mr. Southey's Son-in-Law, the Rev. J. W. WARTER, B.D. New Edition; with medallion Portrait of Southey. Square crown 8vo. 18s. cloth.

SOUTHEY'S THE DOCTOR &c. COMPLETE IN ONE VOLUME.
The DOCTOR &c. By the late ROBERT SOUTHEY. Complete in One Volume. Edited by Mr. Southey's Son-in-Law, the Rev. JOHN WOOD WARTER, B.D. With Portrait, Vignette, Bust of the Author, and coloured Plate. New Edition. Square crown 8vo. 21s. cloth.

ROBERT SOUTHEY'S COMPLETE POETICAL WORKS;
Containing all the Author's last Introductions and Notes. Complete in One Volume, with Portrait and View of the Poet's Residence at Keswick; uniform with Lord Byron's and Moore's Poems. Medium 8vo. 21s. cloth; 42s. bound in morocco.—Or, in 10 vols. fcp. 8vo. with Portrait and 19 Plates, £2. 10s. cloth; morocco, £4. 10s.

SOUTHEY.—SELECT WORKS OF THE BRITISH POETS,
From Chaucer to Lovelace, inclusive. With Biographical Sketches by the late ROBERT SOUTHEY. Medium 8vo. 30s. cloth.

SOUTHEY.—THE LIFE OF WESLEY;

And Rise and Progress of Methodism. By ROBERT SOUTHEY, New Edition, with Notes by the late Samuel Taylor Coleridge, Esq., and Remarks on the Life and Character of John Wesley, by the late Alexander Knox, Esq. Edited by the Author's Son, the Rev. CHARLES CUTHBERT SOUTHEY, M.A. Vicar of Ardleigh. 2 vols. 8vo. with 2 Portraits, 28s. cloth.

STEEL'S SHIPMASTER'S ASSISTANT;

Compiled for the use of Merchants, Owners and Masters of Ships, Officers of Customs, and all Persons connected with Shipping or Commerce; containing the Law and Local Regulations affecting the Ownership, Charge, and Management of Ships and their Cargoes; together with Notices of other Matters, and all necessary Information for Mariners. New Edition, rewritten throughout. Edited by GRAHAM WILLMORE, Esq. M.A. Barrister-at-Law; GEORGE CLEMENTS, of the Customs, London; and WILLIAM TATE, Author of "The Modern Cambist." 8vo. 28s. cloth; or, 29s. bound.

STEPHEN.—ESSAYS IN ECCLESIASTICAL BIOGRAPHY.

From The Edinburgh Review. By the Right Hon. Sir JAMES STEPHEN, K.C.B., Professor of Modern History in the University of Cambridge. Second Edition. 2 vols. 8vo. 24s. cloth.

STOW.—THE TRAINING SYSTEM, THE MORAL TRAINING

SCHOOL, and the NORMAL SEMINARY. By DAVID STOW, Esq. Honorary Secretary to the Glasgow Normal Free Seminary; Author of "Moral Training," &c. 8th Edition, corrected and enlarged; with Plates and Woodcuts. Post 8vo. 6s. cloth.

SWAIN.—ENGLISH MELODIES.

By CHARLES SWAIN, Author of "The Mind, and other Poems." Fcp. 8vo. 6s. cloth; bound in morocco, 12s.

SYMONS.—THE LAW RELATING TO MERCHANT SEAMEN,

Arranged chiefly for the use of Masters and Officers in the British Merchant Service. With an Appendix, containing the Navigation Act; the Mercantile Marine Act, 1850; the general Merchant Seamen's Act; the Seamen's Protection Act; the Notice of Examinations of Masters and Mates; and the Scale of Medicines (Dec. 19, 1850) and Medical Stores (Dec. 20, 1850) issued by the Board of Trade. By EDWARD WILLIAM SYMONS, Chief Clerk of the Thames-Police Court. 4th Edition. 12mo. 5s. cloth.

TATE.—ON THE STRENGTH OF MATERIALS;

Containing various original and useful Formulæ, specially applied to Tubular Bridges, Wrought Iron and Cast Iron Beams, &c. By THOMAS TATE, of Kneller Training College, Twickenham; late Mathematical Professor and Lecturer on Chemistry in the National Society's Training College, Battersea; Author of "Exercises on Mechanics and Natural Philosophy." 8vo. 5s. 6d. cloth.

TAYLER.—MARGARET;

Or, the Pearl. By the Rev. CHARLES B. TAYLER, M.A. Rector of St. Peter's, Chester, Author of "Lady Mary, or Not of the World," &c. New Edition. Fcp. 8vo. 6s. cloth.

TAYLER.—LADY MARY;

Or, Not of the World. By the Rev. CHARLES B. TAYLER, Rector of St. Peter's, Chester; Author of "Margaret, or the Pearl," &c. New Edition; with a Frontispiece engraved by J. ABSOLON. Fcp. 8vo. 6s. 6d. cloth.

TAYLOR.—THE VIRGIN WIDOW:

a Play. By HENRY TAYLOR, Author of "The Statesman," "Philip Van Artevelde," "Edwin the Fair," &c. Fcp. 8vo. 6s. cloth.

TAYLOR.—LOYOLA: AND JESUITISM IN ITS RUDIMENTS.

By ISAAC TAYLOR, Author of "Natural History of Enthusiasm," &c. Post 8vo. 10s. 6d. cloth.

THIRLWALL.—THE HISTORY OF GREECE.

By the Right Rev. the LORD BISHOP OF ST. DAVID's (the Rev. Connop Thirlwall). A New Edition, revised ; with Notes. Vols. I. to V. 8vo. with Maps, 60s. cloth. To be completed in 8 volumes, price 12s. each.   [Vol. VI. nearly ready.

⁎ Also, an Edition in 8 vols. fcp. 8vo. with Vignette Titles, £2. 8s. cloth.

A HISTORY OF GREECE, FROM THE EARLIEST TIMES

to the TAKING of CORINTH by the ROMANS, B.C. 146, mainly based upon Bishop Thirlwall's History of Greece. By Dr. LEONHARD SCHMITZ, F.R.S.E., Rector of the High School of Edinburgh. Second Edition. 12mo. 7s. 6d. cloth.

THOMAS'S INTEREST TABLES.—A NEW SET OF INTEREST

TABLES, from One to Three per Cent. per Annum, calculated by Eighths per Cent. By WILLIAM THOMAS. 4to.   [Nearly ready.

THOMSON'S SEASONS.

Edited by BOLTON CORNEY, Esq. Illustrated with Seventy-seven Designs drawn on Wood, by Members of the Etching Club. Engraved by Thompson and other eminent Engravers. Square crown 8vo. uniform with *Goldsmith's Poems illustrated by the Etching Club*, 21s. cloth ; or, bound in morocco, by Hayday, 36s.

THOMSON'S SEASONS.

Edited, with Notes, Philosophical, Classical, Historical, and Biographical, by ANTHONY TODD THOMSON, M.D. F.L.S. Fcp. 8vo. 7s. 6d. cloth.

THOMSON.—THE DOMESTIC MANAGEMENT OF THE SICK

ROOM, necessary, in Aid of Medical Treatment, for the Cure of Diseases. By ANTHONY TODD THOMSON, M.D. F.L.S. late Professor of Materia Medica and Therapeutics, and of Forensic Medicine, in University College, London. New Edition. Post 8vo. 10s. 6d. cloth.

THOMSON.—TABLES OF INTEREST,

At Three, Four, Four-and-a-Half, and Five per Cent., from One Pound to Ten Thousand, and from One to Three Hundred and Sixty-five Days, in a regular progression of single Days; with Interest at all the above Rates, from One to Twelve Months, and from One to Ten Years. Also, numerous other Tables of Exchanges, Time, and Discounts. By JOHN THOMSON, Accountant. New Edition. 12mo. 8s. bound.

THOMSON.—SCHOOL CHEMISTRY ;

Or, Practical Rudiments of the Science. By ROBERT DUNDAS THOMSON, M.D. Master in Surgery in the University of Glasgow. Fcp. 8vo. with Woodcuts, 7s. cloth.

THE THUMB BIBLE ;

Or, Verbum Sempiternum. By J. TAYLOR. Being an Epitome of the Old and New Testaments in English Verse. A New Edition (1850), printed from the Edition of 1693, by C. Whittingham, Chiswick. 64mo. 1s. 6d. bound and clasped.

TOMLINE.—AN INTRODUCTION TO THE STUDY OF THE

BIBLE: containing Proofs of the Authenticity and Inspiration of the Holy Scriptures; a Summary of the History of the Jews; an Account of the Jewish Sects; and a brief Statement of the Contents of the several Books of the Old and New Testaments. By GEORGE TOMLINE, D.D. F.R.S. New Edition. Fcp. 8vo. 5s. 6d. cloth.

TOOKE.—THE HISTORY OF PRICES :

With reference to the Causes of their principal Variations, from the year 1792 to the year 1838, inclusive. Preceded by a Sketch of the History of the Corn Trade in the last Two Centuries. By THOMAS TOOKE, Esq. F.R.S. 3 vols. 8vo. £2. 8s. cloth.

TOOKE.—THE HISTORY OF PRICES, AND OF THE STATE

of the CIRCULATION from 1839 to 1847, inclusive: with a general Review of the Currency Question, and Remarks on the Operation of the Acts 7 and 8 Vict. c. 32: being a continuation of *The History of Prices, from 1793 to 1838.* By T. Tooke, Esq. F.R.S. 8vo. 18s. cloth.

TOWNSEND.—MODERN STATE TRIALS.

Revised and illustrated with Essays and Notes. By William Charles Townsend, Esq. M.A., Q.C., late Recorder of Macclesfield; Author of " Lives of Twelve Eminent Judges of the Last and of the Present Century," &c. 2 vols. 8vo. 30s. cloth.

TOWNSEND.—THE LIVES OF TWELVE EMINENT JUDGES

of the LAST and of the PRESENT CENTURY. By W. Charles Townsend, Esq. M.A. Q.C. late Recorder of Macclesfield; Author of " Memoirs of the House of Commons." 2 vols. 8vo. 28s. cloth.

TURNER.—THE SACRED HISTORY OF THE WORLD,

Attempted to be Philosophically considered, in a Series of Letters to a Son. By Sharon Turner, F.S.A. and R.A.S.L. New Edition, edited by the Rev. Sydney Turner. 3 vols. post 8vo. 31s. 6d. cloth.

DR. TURTON'S MANUAL OF THE LAND AND FRESH-

WATER SHELLS of the BRITISH ISLANDS. A New Edition, thoroughly revised and with considerable Additions. By John Edward Gray, Keeper of the Zoological Collection in the British Museum. Post 8vo. with Woodcuts, and 12 Coloured Plates 15s. cloth.

TWELVE YEARS AGO: A TALE.

By the Authoress of " Letters to my Unknown Friends," " Some Passages from Modern History," "Discipline," and " Letters on Happiness." Fcp. 8vo. 6s. 6d. cloth.

TWINING.—ON THE PHILOSOPHY OF PAINTING:

A Theoretical and Practical Treatise; comprising Æsthetics in reference to Art—the Application of Rules to Painting—and General Observations on Perspective. By H. Twining, Esq. Imperial 8vo. with numerous Plates and Wood Engravings, 21s. cloth.

TWISS.—THE LETTERS APOSTOLIC OF POPE PIUS IX.

Considered with reference to the Law of England and the Law of Europe. With an Appendix of Documents. By Travers Twiss, D.C.L., of Doctors' Commons; Fellow of University College, Oxford; and Commissary-General of the Diocese of Canterbury. 8vo. 9s. cloth.

URE.—DICTIONARY OF ARTS, MANUFACTURES, & MINES;

Containing a clear Exposition of their Principles and Practice. By Andrew Ure, M.D. F.R.S. M.G.S. M.A.S. Lond.; M. Acad. N.L. Philad.; S. Ph. Soc. N. Germ. Hanov.; Multi. &c. &c. New Edition, corrected. 8vo. with 1,241 Engravings on Wood, 50s. cloth.—Also,

SUPPLEMENT of RECENT IMPROVEMENTS. New Edition. 8vo. with Woodcuts, 14s. cloth.

WATERTON.—ESSAYS ON NATURAL HISTORY,

Chiefly Ornithology. By Charles Waterton, Esq., Author of " Wanderings in South America." With an Autobiography of the Author, and Views of Walton Hall. New Edition. 2 vols. fcp. 8vo. 14s. 6d. cloth.

*** Separately—Vol. I. (First Series), 8s.; Vol. II. (Second Series), 6s. 6d.

ALARIC WATTS'S POETRY AND PAINTING.—LYRICS OF
THE HEART, and other Poems. By ALARIC A. WATTS. With Forty-one highly-finished
Line-Engravings, executed expressly for this work by the most eminent Painters and En-
gravers.

In One Volume, square crown 8vo. price 31s. 6d. boards, or 45s. bound in morocco
by Hayday ; Proof Impressions, 63s. boards.—Plain Proofs, 41 Plates, demy 4to. (only
100 copies printed) £2. 2s. in portfolio ; India Proofs before letters, colombier 4to. (only
50 copies printed), £3. 5s. in portfolio.

WEBSTER.—AN ENCYCLOPÆDIA OF DOMESTIC ECONOMY;
Comprising such subjects as are most immediately connected with Housekeeping : as, The
Construction of Domestic Edifices, with the modes of Warming, Ventilating, and Lighting
them—A description of the various articles of Furniture, with the nature of their Materials—
Duties of Servants, &c. &c. &c. By THOMAS WEBSTER, F.G.S. ; assisted by the late Mrs.
Parkes. New Edition. 8vo. with nearly 1,000 Woodcuts, 50s. cloth.

WESTWOOD. — AN INTRODUCTION TO THE MODERN
CLASSIFICATION OF INSECTS ; founded on the Natural Habits and compounding Organisation of the different Families. By J. O. WESTWOOD, F.L S. &c. &c. &c.   2 vols. 8vo. with
numerous Illustrations, £2. 7s. cloth.

WHEATLEY.—THE ROD AND LINE ;
Or, Practical Hints and Dainty Devices for the sure taking of Trout, Grayling, &c.   By
HEWETT WHEATLEY, Esq. Senior Angler. Fcp. 8vo. with Nine coloured Plates, 10s. 6d. cloth.

WILBERFORCE.—A PRACTICAL VIEW OF THE PREVAIL-
ING RELIGIOUS SYSTEMS of PROFESSED CHRISTIANS, in the Higher and Middle
Classes in this Country, contrasted with Real Christianity. By WM WILBERFORCE, Esq.
M.P.   New Editions. 8vo. 8s. boards ; or fcp. 8vo. 4s. 6d. cloth.

LADY WILLOUGHBY'S DIARY.
So much of the Diary of Lady Willoughby as relates to her Domestic History, and to the
Eventful Reign of King Charles the First, the Protectorate, and the Restoration (1635 to 1663).
Printed, ornamented, and bound in the style of the period to which *The Diary* refers.   New
Edition ; in Two Parts.   Square fcp. 8vo. 8s. each, boards ; or 18s. each, bound in morocco.

YOUATT.—THE HORSE.
By WILLIAM YOUATT.   With a Treatise of Draught.   A New Edition ; with numerous Wood
Engravings, from Designs by William Harvey.   8vo. 10s. cloth.
☞ Messrs. Longman and Co.'s Edition should be ordered.

YOUATT.—THE DOG.
By WILLIAM YOUATT.   A New Edition ; with numerous Wood Engravings, from Designs
by William Harvey.   8vo. 6s. cloth.

₊ The above works, which were prepared under the superintendence of the Society for
the Diffusion of Knowledge, are now published by Messrs. Longman and Co., by assignment
from Mr. Charles Knight.   It is to he observed that the edition of Mr. Youatt's book on the
Horse which Messrs Longman and Co. have purchased from Mr. Knight, is that which was
thoroughly revised by the author, and thereby rendered in many respects a new work.   The
engravings, also, were greatly improved.   Both works are the most complete treatises in the
language on the History, Structure, Diseases, and Management of the Animals of which
they treat.

ZUMPT'S GRAMMAR OF THE LATIN LANGUAGE.
Translated and adapted for the use of English Students, with the Author's sanction and
co-operation, by Dr. L. SCHMITZ, F.R.S.E., Rector of the High School of Edinburgh :
with copious Corrections and Additions communicated to Dr. Schmitz, for the authorised
English Translation, by Professor ZUMPT.   New Edition, corrected.   8vo. 14s. cloth.

[*March* 31, 1851.

WILSON AND OGILVY, SKINNER STREET, SNOWHILL, LONDON.

—

ARDNER'S CABINET CYCLOPÆDI

A SERIES OF ORIGINAL WORKS.

PRICE THREE SHILLINGS AND SIXPENCE EACH VOLUME

The Series, in 132 *Volumes,* £19. 19s.

The LIVES of British DRAMATISTS. By Dr. Dunham, R. Bell, Esq. &c. 2 vols. 7s.

The EARLY WRITERS of GREAT BRITAIN. By Dr. Dunham, R. Bell, Esq. &c. 1 vol.

LIVES of the most Eminent FOREIGN STATESMEN. By G. P. R. James, Esq. and E. E. Crowe, Esq. 5 vols. 17s. 6d.

LIVES of the most Eminent FRENCH WRITERS. By Mrs. Shelley, and others. 2 vols. 7s.

LIVES of the most Eminent LITERARY MEN of ITALY, SPAIN, and PORTUGAL. By Mrs. Shelley, Sir D. Brewster, J. Montgomery, &c. 3 vols. 1s. 6d.

A PRELIMINARY DISCOURSE on the STUDY of NATURAL PHILOSOPHY. By Sir John Herschel. 1 vol. 3s. 6d.

The HISTORY of NATURAL PHILOSOPHY, from the earliest Periods to the present Time. By Baden Powell, A.M. Savilian Professor of Mathematics in the University. 1 vol. 3s. 6d.

... ISE on ARITHMETIC. By D. ... ner, LL.D. F.R.S. 1 vol. 3s. 6d.

A TREATISE on ASTRONOMY. By Sir John Herschel. 1 vol. 3s. 6d.

A TREATISE on MECHANICS. By Capt. Kater and Dr. Lardner. 1 vol. 3s. 6d.

A TREATISE on OPTICS. By Sir D. Brewster, LL.D. F.R.S. &c. 1 vol. 3s. 6d.

A TREATISE on HEAT. By Dr. Lardner. 1 vol. 3s. 6d.

A TREATISE on CHEMISTRY. By Michael Donovan, M.R.I.A. 1 vol. 3s. 6d.

A TREATISE on HYDROSTATICS and PNEUMATICS. By Dr. Lardner. 1 vol. 3s. 6d.

An ESSAY on PROBABILITIES, and on their application to Life Contingencies and Insurance Offices. By Aug. De Morgan, of Trinity College, Cambridge. 1 vol. 3s. 6d.

A TREATISE on GEOMETRY, and its application to the Arts. By Dr. Lardner. 1 vol. 3s. 6d.

A MANUAL of ELECTRICITY, MAGNETISM, and METEOROLOGY. By D. Lardner, D.C.L. F.R.S. &c. and C. V. Walker, Esq. Secretary to the Electrical Society. 2 vols. 7s.

A TREATISE on the MANUFAC- TURE of SILK. By G. R. Porter, Esq. F.R. Author of "The Progress of the Nation," & 1 vol. 3s. 6

A TREATISE on the MANUFA- TURES of PORCELAIN and GLASS. G. R. Porter, Esq. F.R.S. 1 vol. 3s. 6

A TREATISE on the MANUFA- TURES in METAL. By J. Holland, E 3 vols. 10s.

A TREATISE on DOMESTIC ECO- NOMY. By M. Donovan, Esq. M.R.I.A. P fessor of Chemistry to the Company of A thecaries in Ireland. 2 vols.

A PRELIMINARY DISCOURSE on the STUDY of NATURAL HISTORY. By W. Swainson, Esq. F.R.S. L.S. 1 vol. 3s. 6d.

On the HABITS and INSTINCTS of ANIMALS. By William Swainson, Esq. 1 vol. 3s. 6d.

A TREATISE on the NATURAL HIS- TORY & CLASSIFICATION of ANIMALS. By W. Swainson, Esq. 1 vol. 3s. 6d.

On the NATURAL HISTORY and CLASSIFICATION of QUADRUPEDS. By W. Swainson, Esq. 1 vol. 3s. 6d.

On the NATURAL HISTORY and CLASSIFICATION of BIRDS. By W. Swainson, Esq. 2 vols. 7s.

ANIMALS in MENAGERIES. By W. Swainson, Esq. 1 vol. 3s. 6d.

On the NATURAL HISTORY and CLASSIFICATION of FISH, REPTILES, &c. By W. Swainson, Esq. 2 vols. 7s.

The HISTORY and NATURAL AR- RANGEMENT of INSECTS. By W. Swain- son, Esq. and W. E. Shuckard, Esq. 1 vol. 3s. 6d.

A TREATISE on MALACOLOGY; or, the Natural Classification of Shells and Shell-Fish. By William Swainson, Esq. 1 vol. 3s. 6d.

TREATISE on TAXIDERMY. With the Biography of the Zoologists, and Notices of their Works. 1 vol. 3s. 6d.

A TREATISE on GEOLOGY. By John Phillips, F.R.S. G.S. Professor of Geo- logy, King's College, London. 2 vols. 7s.

The PRINCIPLES of DESCRIPTIVE and PHYSIOLOGICAL BOTANY. By the Rev. J. S. Henslow, M.A. &c. 1 vol. 3s. 6d.

DON: LONGMAN, BROWN, GREEN, AND LONGMAN

CPSIA information can be obtained
at www.ICGtesting.com
Printed in the USA
BVHW071439140819
555860BV00025B/2122/P